Impact on Ethylene Propylene Diene Rubber Particles on Nano Resins

Sanjeev Kumar

TABLE OF CONTENTS

LIST OF FIGURES

LIST OF TABLES

ABSTRACT

Elastomeric materials have an important function in nuclear fuel reprocessing plants as o-rings, gaskets, airlock doors, hatches, master slave manipulators, radiation shields, etc. These components have to withstand the combined effect of gamma radiation and hydrocarbon solvents. Ethylene Propylene Diene Monomer (EPDM) rubber is the material of choice for elastomeric components in nuclear applications due to its radiations resistance. Though radiation resistance of EPDM is remarkable compared to other rubbers, its ability to endure hydrocarbon solvents used in reprocessing is low. To enhance the durability of EPDM in such environments, EPDM - Chlorobutyl rubber (CIIR) blends of varying compositions were developed and characterized for mechanical, thermal, and solvent sorption behavior. Spectroscopic and morphological analyses were used to evaluate the compatibility of blends. Due to synergistic effect, the optimal composition of blends with superior mechanical properties and solvent resistance were found to be 60-80% EPDM and 20-40% CIIR. The effect of gamma irradiation at three different cumulative doses (0.5, 1 and 2MGy) from ^{60}Co source was studied on the optimized blends. Based on spectroscopic, morphological, mechanical, thermogravimetric and sorption properties of irradiated blends, the blend containing 80% EPDM was found to have superior retention of properties after irradiation.

The blends were reinforced with organo-modified layered silicate (nanoclay) to further enhance their performance in radiation as well as hydrocarbons environments. The effect of nanoclay or layered silicates on mechanical and viscoelastic properties of EPDM-CIIR blends were studied in this thesis. The mechanical properties of the nanocomposites increased (upto 57 %) and solvent transport coefficients decreased (by 30%) with increasing nanoclay content. The morphology and physico-chemical interactions were evaluated by XRD, TEM and FTIR and correlated to the enhancement in mechanical properties. Mooney Rivlin plots provided insight into the non-linear mechanical behavior of EPDM-CIIR nanocomposites. From dynamic mechanical analysis (DMA), it was found that blend with 5 phr layered silicate content had the maximum storage modulus. Significant lowering as well as broadening of tanδ peak was observed for the blend with well dispersed nanoclay (5 phr). The entanglement density and constrained volume near the interfaces, calculated from DMA data, gave into the reinforcing mechanism. Rheological characteristics of the nano-reinforced blends also revealed stiffening effect of layered silicates.

Payne effect and stress relaxation studies confirmed good rubber-nanoclay interactions in the nanocomposites. The applicability of various analytical models to predict the static and dynamic modulus as well as Payne effect was explored in this thesis. The effect of nanofiller content on the mechanical properties, solvent uptake and thermal degradation of blends exposed to gamma radiation was also investigated by irradiating the nanocomposites with gamma rays for cumulative doses up to 2MGy. Depending on the dose of cumulative radiation exposure, both chain scission and / or cross-linking occurred in the nanocomposites resulting in varying degrees of changes in properties. The chemical changes due to radiation induced effects were evident from FTIR analysis.

The influence of bis (3-triethoxysilylpropyl) tetrasulfide (TESPT) grafted nanosilica particle reinforcement on the mechanical, viscoelastic, thermal and transport characteristics as well as behavior after exposure to different cumulative γ-radiation doses of EPDM-CIIR blends is detailed in this study. The tensile strength and modulus of the nanosilica reinforced were enhanced upto 64% and 118% respectively whereas solvent diffusion coefficient reduced by 22%. Enhancement in thermal stability, viscoelastic and barrier properties were observed after reinforcement. γ-radiation ageing resistance of EPDM-CIIR blends improved with incorporation of nanosilica, with blends containing 7.5phr NS showing optimum properties and radiation ageing resistance. The larger interface provides effective stress transfer and creates barrier to free radical and solvent permeation. The applicability of Korsmeyer-Peppas, Peppas-Sahlin and Higuchi models to predict of sorption behavior are investigated. Coats-Redfern and Horowitz-Metzger models were employed to evaluate the activation energy for thermal degradation. TEM, FTIR and rheological curves were utilized to corroborate improvement in mechanical and solvent sorption behavior. FTIR and ESR analysis provided insight on the chemical changes in the nanosilica reinforced blends after irradiation.

Preliminary studies on the effect of MWCNT reinforcement on EPDM based blends were also carried out to evaluate the improvement in mechanical properties as well as hydrocarbon and gamma radiation resistance. The improvements in properties were correlated to dispersion and interfacial interactions of MWCNT, which were confirmed by spectroscopic and morphological analysis. MWCNT reinforcement reduced the magnitude of changes in mechanical and transport properties after γ-irradiation. ESR and FTIR spectra provided qualitative information on free radical formation and chemical changes due to γ-rays exposure. This study also aims to develop a carbon black-nanofiller hybrid composite of EPDM/CIIR blends for product application. Hybrid nanocomposites were prepared by reinforcing EPDM blends with varying amount of nanoclay and nanosilica along with carbon black. The synergistic effect of hybrid fillers together with tortuous path decreased solvent permeation and free radical migration resulting in reduction of the radiation ageing effects in the hybrid nanocomposites, making it potential and suitable candidature for application in nuclear fuel reprocessing facilities.

Chapter 1

Introduction and literature review

1.1. Nuclear fuel reprocessing

The reprocessing of spent nuclear fuels involves recovery of fissionable material which is capable of sustaining a nuclear fission reaction [1]. The spent nuclear fuel reprocessing is used to extract fissionable remains of uranium and plutonium for recycling and eliminating wastage and emission of high-level radioactive toxins from nuclear fuels to the environment. This contributes towards saving national nuclear energy reserves and reducing the volume of toxic nuclear waste to about one-fifth [2], [3]. From the historical perspective, the reprocessing of nuclear fuels emerged when workers in early period recognized the possibility of feasible chemical separation of plutonium and isotopic separation of enriched uranium. Spent fuel reprocessing differs from conventional chemical processing due to the presence of radioactive materials. The generation of nuclear power from controlled nuclear fission chain reactions of heavy elements is the most significant technical application of nuclear reactions in the present scenario. The world's reserves of energy in nuclear fuels are emerging as an intense and highly demanded energy producing arena because the depletion of energy reserves based on gas, oil and coal are occurring at a faster pace.

NUCLEAR FUEL CYCLE

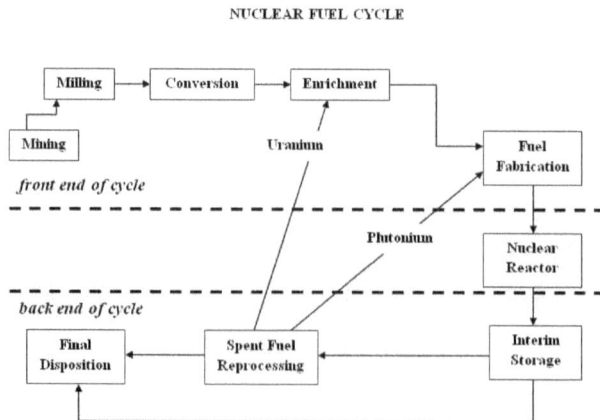

Figure 1.1. Nuclear fuel cycle [4]

1

The initial development of nuclear fuel reprocessing was based on a precipitation process known as "bismuth process". In this method, several series of precipitation processes were performed to extract plutonium remains, but the drawback was that it was unable to recover uranium extracts and large amount of radioactive wastes were eliminated. The replacement of bismuth process by solvent extraction method is used presently to recover both uranium and plutonium. The PUREX (Plutonium Uranium Extraction) [5] in late 1950s, is generally based on solvent-solvent extraction method which uses paraffinic hybrocarbon solvents like tri-n-butyl phosphate (TBP) diluted with dodecane solution as a solvent to extract and separate elements from radioactive products [1]. The polymer based materials used in the nuclear power plants are subjected to radiation and corrosive environments [6], [7], [8], [9].

1.2. Elastomers used in nuclear fields

Elastomeric materials have an important function in nuclear fuel reprocessing plants as o-rings, gaskets, airlock beadings, hatches, belts, glove boxes, shielding elements, etc. These components are subjected to withstand intense radioactive environments. In the last decade, interest has been drawn to study of elastomeric materials in nuclear applications [10]. Elastomeric materials in nuclear fuel reprocessing plants resist spillage of radioactive emanations from spent nuclear fuels. The radioactive emanations from spent nuclear fuels are alpha, beta, gamma and neutron [6], [9], high active streams will be having bulk of gamma radiations. In this context, lot of technical research is required for attaining longevity of elastomeric materials in resisting gamma radiations. The stability of elastomers against gamma radiation doses are evaluated to recommend their applications in nuclear fields [11].

The longevity of elastomers used in nuclear fields should be more considering the safety aspects during replacement and costs of breakdown as well as maintenance [12] [13]. The criteria for selection of elastomers in nuclear applications involve mechanical properties, ageing and degradation phenomena when subjected to radioactive environments [14].

Elastomers are used for several intense applications at moderate temperature due to their fundamental viscoelastic behavior, conformability, resilience, toughness, mechanical properties etc. The base rubber is compounded with compounding agents like activators, accelerators, anti-

2

oxidants, varying amounts of fillers and curing agents to meet functional properties for end applications [10].

1.3. Radiation and its interactions with materials

The radioactive emanations from spent nuclear fuel typically comprise of high energy radiations like alpha, neutrons, beta and gamma rays. The radiation interaction with the materials plays a vital role in the selection of materials for nuclear fuel reprocessing applications [15]. The incident radiations lose their energies when interactions occur through ionization and non-ionization processes. The dosage of irradiation is determined on the basis of energy absorbed by the material and the SI unit is Gray (Gy), where 1Gy equals to 1J/kg of matter absorbed [16]. The ionizing radiation possesses high energy to eject electrons from atoms, generating ions and charge distributions and also excites the electron from ground state to excited state. The ionizing radiations viz. gamma rays, x-rays, etc., occur in either wave form or particle form. Electromagnetic radiations viz. x-rays and gamma rays lose their energies in three ways while passing through matter: pair production, compton scattering and photoelectric absorption [15].

1.3.1. Pair production

In pair production, x-ray or gamma ray photon produces a positron-electron pair within the medium. It is necessary for the photon energy to be higher than the equivalent of the remaining masses of two particles, i.e. more than 1MeV. The electron-positron pair is produced by the disappearing electrons from the target. The positron interacts with another electron and generates two gamma photons of energy 0.5MeV (annihilation radiation) whereas the positive ion absorbs electron. Both electrons and positrons produced are energetic and lose their energies by ionization and excitation [17].

1.3.2. Compton scattering

The photons lose part of their energy by ejecting electrons from atoms in compton scattering. The electrons cause ionization and excitation while the scattered photons of reduced energy interact further, either by undergoing compton scattering or by photoelectric absorption. Compton scattering is a complicated function of radiation energy. Elastic scattering or Rayleigh scattering is one of the radiation interactions without energy absorption [17].

3

1.3.3. Photoelectric absorption

The photon is absorbed by an atom with the ejection of a fast electron, usually from one of the inner shells in photoelectric absorption. The energy of the electron is the difference of energy of incident photon and binding energy. The binding energy is in the form of an x-ray, which in turn undergoes photoelectric absorption, but is more likely to be used to eject another electron from the same atom. This latter process is known as the Auger effect and the ejected electron is called an Auger electron. The ionized atom returns to the ground state by the ejection of Auger electron with some characteristic radiations [15]. The energy of the orbital electron (E_e^-) ejected is calculated by the following relation:

$$E_e^- = E_\gamma - B.E_e$$

where $B.E_e$ is the binding energy and E_γ is the incident photon energy.

1.4. Effect of gamma radiation on elastomers

The study of the physical properties of elastomeric materials subjected to high-energy radiations has attracted considerable attention since 1940s. The knowledge about changes in the properties of a polymeric substance has a significant role in selection of components used in nuclear fuel reprocessing facilities and for handling radioactive materials. The changes caused by radiation are derived from radiation induced effects. In the case of radiation induced changes the factors like dosage rate, temperature and environment are also important. As discussed in the previous section, when high energy ionizing radiation passes through matter, large fraction of incident energy is dissipated through interactions either with the nucleus or with the orbital electrons. As a result of radiation interactions on organic polymers, reactive species like ions and radicals are formed. The formations of these reactive species are associated with rupture of covalent bonds in the polymer. These species will readily react with each other or with the environmental oxygen and new oxygenated chemical bonds are formed in the elastomer. The existence of new bonds alters the structure of the elastomeric material, resulting in changes in its physical and chemical properties. The whole effect of interaction process (radiations) occurring by radical mechanism in organic polymeric materials is explained in detail below

4

Ionization[18]

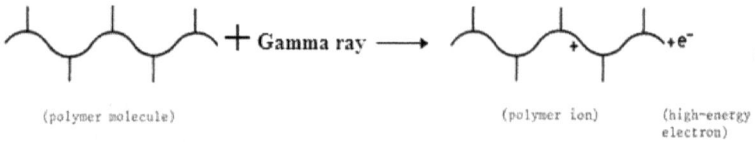

(polymer molecule)   (polymer ion)   (high-energy electron)

Figure 1.2 (a) Ionization [18]

Ionization occurs if the transfer of energy in the interaction is greater than the binding energy of an electron in its parent molecule.

Excitation [18]

Excited polymer molecules are produced either by the direct excitation due to weak interactions, viz., fast moving charged particles, or as a result of ion-thermal electron recombinations.

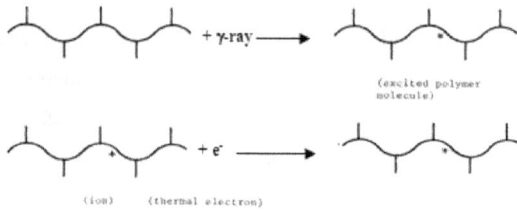

(excited polymer molecule)

(ion)   (thermal electron)

Figure 1.2 (b) Excitation [18]

Dissociation [18]

When the energy gained by the polymeric molecules becomes equal to ionization energy, the dissociation or rupture of covalent bonds with bond energies of the order of 5 eV occurs within the polymer molecule as shown in the schematic below. In elastomeric substances, the broken bonds can occur either in the main chain or between the main chain and the side chain.

5

i) Main-chain rupture

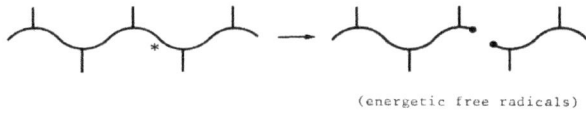

(energetic free radicals)

ii) Side-chain rupture

Figure 1.2 (c) Dissociation [18]

The most important chemical changes that occur during irradiation of polymers are cross-linking and chain scission or degradation. When an elastomer is irradiated these processes usually occur competitively but at different rates. The free radicals also interact with atmospheric oxygen to produce oxygenative degradation of rubbers [18].

Cross-linking [18]

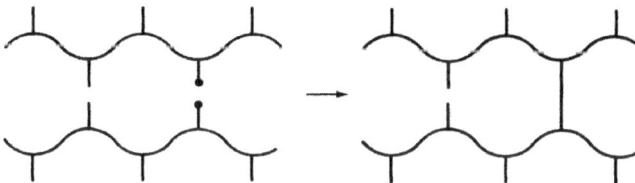

Figure 1.2 (d) Crosslinking [18]

The radicals interact with each other by forming chemical bonds between polymeric chains

Chain scission with oxidation [18]

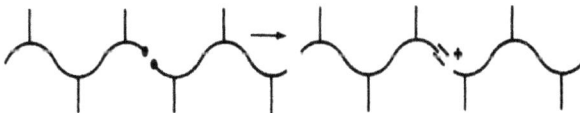

6

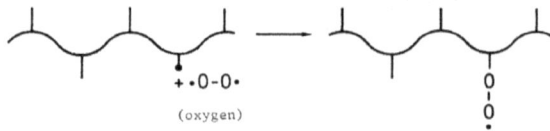

Figure 1.2 (e) Chain scission with oxidation [18]

The radicals interact with the atmospheric oxygen to form peroxide radicals and also fracture the polymer chains. The formation of free radicals is the key factor for the degradation of material properties in irradiated elastomers [19]. As explained above, free radicals are initiated at the backbone chain of an elastomeric or a polymeric structure when it is exposed to an ionizing radiation, high energy particles or photon. The presence of fast and slow decaying free radicals was found when elastomers were gamma irradiated from Co^{60} source. It was revealed that fast decaying radicals were mostly from alkyl type radicals residing in regions of low crosslink density and long-lived radicals were attributed to radicals residing in highly crosslinked region of the polymer structure. Several literatures are available stating about the significance of organic polymeric and elastomeric materials in nuclear technologies [20]. They rank with metallic and ceramic materials as the important category of materials for components used in nuclear fields [20].

The large scale use of elastomers in nuclear power plants, high energy particle accelerators and industrial as well as research irradiation facilities motivates extensive investigations on radiation induced degradation [21]. The effects of radiation induced ageing are predominantly found on intermolecular chemical changes, macromolecule scission and formation and/or consumption of unsaturated structure. The modifications are induced by high energy radiations like accelerated electrons, gamma rays, etc. based on the ability of the compound to produce free radicals. High concentration of radicals is easily produced in branched polymers or polymers with unsaturated bonds than on saturated polymers. Crosslinking is promoted by high energy radiation depending upon the rate of free radical formation [22] [23]. Some reports are also available on the studies of radiochemical effects of elastomer blends evaluated by isothermal chemiluminescence methods [24]. In another study by K.S. Seo *et al.* [25], radical decay behavior was evaluated with the aid of ESR spectroscopy.

7

Most common elastomers used in nuclear plants are Ethyelene Propylene Diene Monomer (EPDM), Nitrile rubber (NBR), Fluroelastomer (FKM, Viton), Polychloroprene rubber etc. Nowadays, the selection criteria for elastomeric materials in nuclear fields have attained wider importance.

1.5. EPDM and effects of radiation on EPDM rubbers

EPDM was commercially introduced in the late 1960s and is the fastest growing synthetic rubber for many applications. Ethylene and propylene are the main components of EPDM rubber. The distinguishable chemical structure of EPDM makes it to serve for both general purposes as well as for special purpose applications. EPDM elastomer has several uses in specialty applications like nuclear technology, aerospace, automobile sectors, radiation shields, etc. due to its low specific density, cost, ease of processability and weatherability. Apart from these, superior resistance to ozone, radiation, weathering and other environmental conditions viz. heat, light, oxygen, etc. led to the utility of EPDM in intense applications in nuclear fields. EPDM has a prominent position amongst elastomers for upcoming technologies due to its excellent heat resistance and low density. The structural design of EPDM is capable of accepting the reinforcement by fillers which open a wide range of industrial applications. The unsaturated chemical structure of Ethylene Propylene Monomer (EPM) is associated with a non-conjugated diene in the same group. The reactive unsaturation sites for sulphur curing or polymer modification are provided by the small amount of non-conjugated diene monomers placed at side chain of the back bone structure. The dienes are structured such that only one of the double bonds polymerizes. The dienes can be hexadiene (HD), ethydiene norborene (ENB) and dicyclopentadiene (DCPD). These dienes are polymerized with EPM depending upon the applications to produce unsaturated EPDM. ENB is mostly used as diene as it can be incorporated easily and has greater reactivity with sulfur cross-linking. EPDM with ENB is chosen in the present study and its chemical structure is shown in Figure 1.3 below. The properties of EPDM elastomer vary with the percentage composition of Ethylene and Propylene contents in the chemical formulae.

8

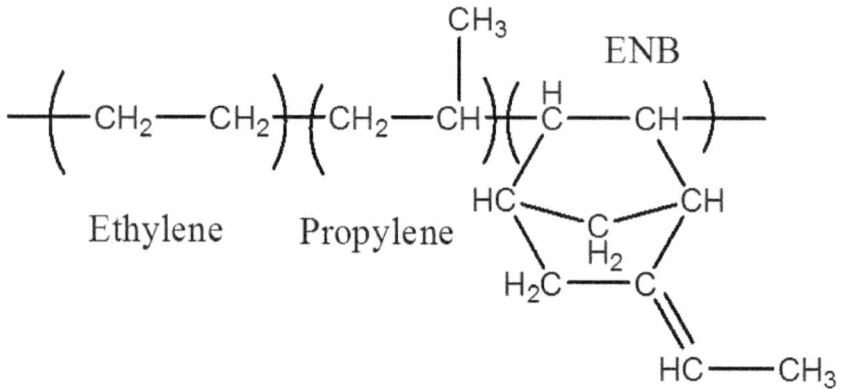

Figure 1.3. Chemical structure of EPDM rubber [23]

Glen Lee [26] has carried out a comparative study on radiation resistance properties of elastomers. It was revealed that EPDM compounds were found to exhibit acceptable properties for O-rings after irradiating at 5×10^8 rads while Viton failed at 1×10^7 rads. EPDM compounds showed excellent all round properties, viz., tensile strength, compression set and hardness with radiation levels up to 5×10^8 rads when compared to sulfur cured urethane, which showed less compression strength and elongation after radiation level of 10^8 rads [26]. Other rubber compounds like NBR and Neoprene became brittle after radiation level of 3×10^8 rads. EPDM based compounds (559N and 559EQ) retained reasonable flexibility, strength, hardness and very good compression set resilience. It was concluded that EPDM rubbers were the best for all radiation resistance applications up to 5×10^8 rads. It was also evident that EPDM rubbers can be used even for radiation dose higher than 5×10^8 rads provided some difficulties would be encountered in designing a functional part because of its high compression set property [26].

Zenoni *et al.* [27] presented a comparative study of four compounds one based on fluroelastomer polymer (FPM) and three based on EPDM to understand their radiation resistance behavior in mixed neutron and gamma fields with high dose levels. It was observed that EPDM based compounds exhibited more stable behavior after irradiating at accelerated dosage of 2MGy. The FPM elastomer, in spite of its lower dose absorption in neutron fields, showed increase in stiffness and brittleness with the largest variation in properties compared to EPDM. FPM and

EPDM elastomers are employed for application as o-rings and gaskets in combined neutron and gamma radiation ageing fields due to the ability to withstand radiation and high temperatures. EPDM having larger hydrogen content absorb higher amount of radiation dose when placed in neutron or gamma fields than FPM [28].

Several researchers proved the utility of EPDM seals in nuclear power plants by carrying out accelerated ageing tests on regular service conditions [29]. F. Le Lay [30] studied the combined effect of gamma radiations at 10, 20 and 400kGy and thermal ageing of EPDM seals for its utility in nuclear powered vessels for a service period of upto 32 years. EPDM rubber, known for its enhanced radiation and thermal resistance properties were chosen over NBR, styrene butadiene rubber (SBR) and fluorocarbon elastomer (FKM) for seal applications exposed to combined ageing environments like temperature, radiations, oil, water, oxygen etc. in nuclear power vessels. A satisfactory correlation was obtained between physicochemical and mechanical properties of EPDM seals.

Sandra *et al.* [31] studied the influence of gamma radiation doses of 25kGy, 50kGy, 75kGy, 100kGy and 200kGy on EPDM rubbers used in electric cables and wires in nuclear plants. It was observed that the "tensile strength" enhanced till 25kGy as an effect of radiation induced crosslinking and started decreasing due to the dominance of chain scission. It was concluded that EPDM was most radiation resistant amongst other rubbers due to the presence of large number of hydrogen atoms and saturated chemical bonds on its principal backbone chemical structure [31]. From a comparative study on EPDM and Natural rubber (NR) by Bayram *et al.* [32], it was revealed that excellent thermal and ozone resistance of EPDM due to its saturated chemical structure can replace NR for dynamic applications. The concluding remarks were the following: rebound resilience was superior in EPDM by 15% than NR and heat buildup was lower by 20% than NR.

Nakano *et al.* [33] have carried out tests on radiation resistance behavior of several elastomers like EPDM, silicone rubber and fluoro-rubber for gasket applications. EPDM compounds produced good compression set results after gamma radiation exposure when compared to silicone rubber. The mechanical property of fluoro-rubber also deteriorated compared to EPDM rubber when subjected to radiations. EPDM exhibited lower compression set at high radiation dose of over 300kGy. The selection criterion of gaskets in nuclear applications is dependent on

10

resistance of rubber against thermal and radiation ageing. The decomposition rate of flurorubber was very high compared to EPDM rubber after irradiation. The reason for enhanced radiation resistance characteristics in EPDM rubber can be attributed to the fact that the saturated chemical structure can absorb more incident energy of radiation compared to other rubbers. The extraction rate tests were carried out on the irradiated rubber samples by the above researchers to evaluate release rate of rubber decomposition caused by radiation. It was observed that EPDM rubbers had the least decomposition rate whereas fluororubber decomposed at a faster pace after irradiation. This hypothesis was also validated by the variation in chemical changes evaluated by FTIR spectroscopy.

In a similar study by the same research group [34], the variations in the chemical structure of rubbers, when subjected to severe gamma ray dosage of 3MGy and for 400h exposure time, were reported. The researchers determined that EPDM rubbers were more suitable for manufacturing high radiation resistant gaskets. From the test results, it was also evident that EPDM rubbers can incorporate more inorganic additives or fillers than silicone rubber and FKM because of their low viscosity. FKM has higher viscosity and cannot attain much reinforcing effect.

Generally, an elastomeric substance exhibits a wide range of radiation effects. The formation of bonds (crosslinks) or breakage of bonds (chain scission) after irradiation is manifested as change in chemical and physical structures, appearances, mechanical properties etc. The radiation stability is dependent on the chemical structure of the material because radiation-induced changes are often localized at a specific bond rather than coupled to the entire chemical structure [35].

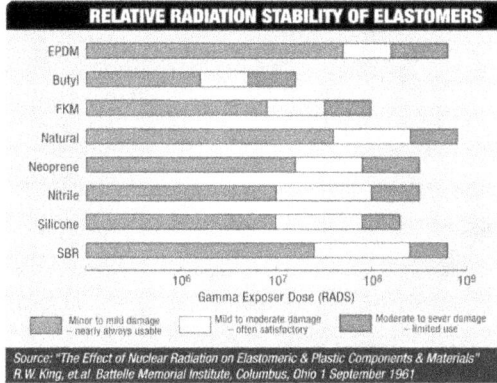

Figure 1.4. Relative radiation stability of elastomers [36]

Variety and features of rubbers offering radiation resistance

Brand name	Type	Features	Permitted Thermal Condition (rough standard)	Confirmed Permitted Radiation Level
100 series	EPDM base	High mechanical strength	-35 to 90°C	8 MGy
300 series	EPDM base	Wide range of application	-35 to 90°C	20 MGy
500 series	EPDM base	• Suitable for high vacuum material • Out-gassing performance equivalent to or higher than that of fluoro rubber (FKM)	-50 to 100°C	1.2 MGy
700 series	Liquid BR	Two-pack room temperature hardening type liquid sealant	-20 to 70°C	2.6 MGy
900 series	Synthetic Rubbers	Suitable for heat and oil resistant element	-20 to 130°C	2 MGy

※Please note that radiation resistance characteristics may vary depending on the evaluation method.

Figure 1.5. Features of radiation resistance rubbers [31]

1.6. Review of EPDM rubbers

Apart from radiation resistance, EPDM rubbers have good resilience and tear resistance, making it useful for many engineering applications. Their thermal and electrical insulation properties are also rated from good to excellent. EPDM is used for applications involving thermal shields because of its thermal insulation behavior [29], [31]. This is also utilized for insulation purposes in high voltage cables due to its excellent electrical insulation characteristics. The hydrocarbon

12

chain of EPDM elastomer is environmental friendly and it does not possess any toxic components. Its low coefficient of thermal expansion and good ablation properties facilitates extra advantage owing to the use of this elastomer in many critical environmental conditions.

Zaharescu et al. [37] studied oxygenative degradation of gamma irradiated ethylene propylene elastomers (EPR and EPDM) in the presence of antioxidants. The stabilization effect of antioxidants modifies the distribution of oxygenated compounds [38]. The rate of degradation is dependent on elastomer matrix, antioxidant chemistry and radiation energy absorbed by the rubber. From a study by Rivaton et al. [39] it was revealed that oxidation rate in EPDM rubbers can be effectively reduced by antioxidants in thermo-oxidative conditions. Further, effective stabilization can be obtained by incorporating both photo oxidation and radio oxidation stabilizers. Radiation resistance depends on the formulation of rubbers, conditions of radiation exposure such as atmosphere, temperature dose rate, mechanical stress, etc.[40]. Amongst these, oxidation induced by the radiation is the most influencing [41].

Worawat et al. [42] developed EPDM rubber composites for gamma ray shielding applications. The gamma ray shielding properties of EPDM rubber increased for composites made of EPDM and metal oxides. EPDM composites are used for development of polymer based radiation shield applications [43], [44]. The reason for using EPDM elastomers for gamma ray shielding applications is its lower density of about 0.86-0.90 kg/m^3 which can take up more fillers [45]. It was stated that the reinforcement of a polymer by nano and micro fillers enhanced radiation resistance [46]. T Zaharescu et al. [47] also carried out studies on radiation stability of ethyelene propylene terpolymers subjected to gamma radiation. The efficiency of two phenolic antioxidants was assessed for the qualification of EPDM as high stabilized product in radiation field. The chemical changes were revealed from changes in FTIR spectra, which were correlated to mechanical testing and chemiluminescence. The qualification tests were carried out at dose rate of 0.04 and 0.4 kGy/hr for simulating real parameters of radiation aged materials. The improvement in thermal and radiation stability were provided by saturated chemical structure of EPDM which absorbs incident energy and functioning of phenolic anti-oxidant function which interacted with the free radicals [47]. The behavior of elastomers under irradiation was found to be dependent on the additives in rubber formulation as well as irradiation dose, dose rate, and environmental conditions [48].

13

Yuetao *et al.* [49] studied the effects of gamma ray radiation on the properties of flurosilicone rubber (FSR). The morphology, static and dynamic mechanical properties as well nuclear magnetic spin resonance spin-spin relaxation analysis were examined before and after irradiation. Though the change in properties of FSR ascribed to both degradation and crosslinking, the overall results revealed dominance of degradation reaction over crosslinking during irradiation. The decline in storage modulus of sample after irradiation from dynamic mechanical analysis indicated occurrence of degradation or chain scission during gamma irradiation. The mechanical properties like tensile and tear strength as well as elongation at break decreased over investigated range of radiation dose.

1.7. Assessment of EPDM based blends in nuclear fuel reprocessing facilities

Elastomers selected for applications in nuclear fuel reprocessing facilities are required to possess both radiation and hydrocarbon solvent resistance. From the preliminary literature survey in the above sections, it is clear that EPDM rubber is the material of choice for O-rings, gaskets, seals, master slave manipulators, gamma radiation shields, etc. in radioactive environments due to its excellent resistance towards heat, ozone, environmental degradation and gamma radiation. In nuclear reprocessing plants, most of the elastomeric components are exposed to paraffinic hydrocarbon solvents in addition to radiation [14] [9].

Though the radiation resistance of EPDM is remarkable compared to that of other elastomers, its ability to endure paraffinic hydrocarbon solvents like tributyl phosphate (TBP) and dodecane used in reprocessing environments is low. The shortcomings of EPDM rubber in enduring hydrocarbon environment can be overcome by several ways. One method is to blend EPDM with a suitable rubber having hydrocarbon resistance. Another method is to compound it with an additive that improves its solvent barrier properties. Blending of rubbers is an economically viable, well known and versatile method to prepare latest engineering materials. Blending of two or more rubbers enables to overcome certain deficiencies of one rubber by the advantages of its counterpart. It develops a new elastomer with a novel structural system with a combination of both copolymers. Blending of elastomers or polymers [50] also leads to the development of a new property or enhancement of exiting characteristics of an organic material to meet any specific requirements. The miscibility in elastomer blends is produced as a result of specific interactions between the blend components [51] [52]. These interactions include hydrogen

14

bonding, van der waals forces, dipole-dipole interactions, etc. Certain blends are miscible even in the absence of any specific interactions [51].

In general, blending enhances the mechanical properties, ageing resistance and processing characteristics. Several studies have been reported in literature review to achieve better physical properties, improve elastomeric properties, achieve better processing behavior, and to attain lower cost by blending two polymers. The effects of blend ratios and crosslinkages have shown to be significant in achieving specific improvements in the properties of blends [53] [54]. Investigations on the effect of blend ratios and crosslinking on mechanical and ageing behavior of EPDM–styrene butadiene rubber (SBR) blends have been carried out by Nair *et al.* [53]. Compatibility study has been reported on blends with natural rubber and EPDM by EI-Sabbagh [55]. Solvent resistance and mechanical property evaluation of thermoplastic elastomeric blends of chlorobutyl and nitrile rubber has also been reported by some researchers [56]. Jose *et al.* [57] has reported that blending EPDM with precured CIIR results in improved mechanical properties and compatibility.

Hassan *et al.* [58] investigated the effect of gamma irradiation resistance and nanoclay content on dye sorption of EPDM rubber composites. In devulcanized EPDM–polypropylene blends exposed to radiation, a drastic deterioration in physical properties was observed due to chemical degradation of polymer chains caused by high energy radiation that produce chain scission, chain branching, crosslinking, and possible oxidation [58]. The role of blend morphology on Styrene Butadiene Rubber (SBR)/EPDM rubber blends was reported by K.A. Dubey *et al.* [59]. The enhancement in miscibility of blends was observed after subjecting to gamma radiation. The experimental and theoretical value of intrinsic viscosity was compared with the weight percentage of EPDM content. The theoretical values were calculated on the basis of ideal behavior assumption. It was concluded that the miscibility of SBR-EPDM blends was compatible for composition range 0 to 60% EPDM content when it is pre irradiated at an absorbed dose of 10kGy. Additionally, enhancement in adhesion properties, tear strength, good weatherability and hardness were also obtained on the blends prepared.

The properties of EPDM/Acrylonitrile Butadiene rubber (NBR) blends made using sulfur cross linking were studied on the basis of blend ratio by some researchers [60] and attempts were made to correlate mechanical properties with analytical models. Due to synergistic effects of both

15

rubbers, tensile properties like tensile strength and tear resistance augmented by 30% for blends with 70% EPDM rubber content. The applicability of theoretical models like upper bound (parallel), lower bound (series), Halpin Tsai and Maxwell models was analyzed to predict mechanical behavior of EPDM/NBR blends. The experimental values of tensile properties showed significant positive deviations from the values predicted from upper and lower bound models suggesting better properties for the blend system [60] [61].

The blending of EPDM with another rubber having barrier resistance to long chain hydrocarbon solvents would meet the specific requirement of withstanding both radiation and hydrocarbon environments simultaneously. Chlorobutyl rubber (CIIR) is a promising candidate for blending with EPDM. The polar chlorine group in CIIR renders hydrocarbon resistance to it. CIIR is an elastomeric isobutylene isoprene copolymer containing reactive chlorine atoms. The chlorination of isobutylene isoprene rubber (IIR) was first emphasized by Exxon workers during 1950s and the first commercial product was introduced in 1960 [62]. Chlorination of IIR was done by substitution process by free radical mechanism. Halogenation (Chlorination or Bromination) behavior of IIR has provided a better insight on the characteristics of IIR. The presence of methyl group in the reactive site on the structure of IIR profoundly influenced the possibility of halogenation. The product after chlorination of IIR was distinct from other typical patterns of other tri substituted alkenes due to steric hindrance imposed by halogen groups. The absence of chlorine products across the double bond demonstrated this distinction [62]. This model compounding approach adopted on halogenations of IIR by Vukov R [62] provides a valuable tool for fundamental understanding of IIR system.

S.H. Botros [63] reported in detail on the thermal stability of IIR/EPDM blend compared to IIR rubber. The drawback of heat, ozone and marginal green strength of IIR were surmounted by blending with various types and ratios of EPDM having excellent resistance to ozone and radiations. The physio-mechancial properties of IIR/EPDM blends before and after thermal ageing at various temperatures were noticed for conveyer belt applications that involve exposure to high temperatures. Blends of IIR and Keltan-820 type EPDM rubbers produced enhancement in thermal resistance as the ageing temperature rises up to 120-130°C. IIR:Keltan-820 grade EPDM (70:30) blend exhibited outstanding resistance to thermal resistance since it could retain 100% of its original strength after exposure to ageing conditions at 165°C for 48h. It was also elucidated that resin cured IIR produced similar physico-mechanical characteristics as that of sulphur cured IIR though the former cured rubber was characterized by lower cure and scorch times as well as low maximum torque.

In another literature [64], the crack-growth resistance behavior of EPDM blended with Brominated butyl rubber (BIIR) under dynamic loading conditions over a temperature range was investigated. It was revealed that EPDM:BIIR at 30:70 produced best fatigue resistance due to its morphology and strain energy density factors. The crack-growth resistance escalated with increase in BIIR content. In the gum and blend with higher EPDM content, crack growth propagation was faster. The fractures surfaces were examined by microscopic techniques and it was noticed that blends exhibited fracture features intermediate of its constituent rubbers. The reduction in fatigue life was observed at higher temperature (100°C).

IIR/EPDM rubber blends for inner tube applications were reported by Thomas and David [65]. The partial replacement of IIR has minimized softening of IIR upon oxidation occurring at inner tube while running at high temperatures and load conditions. The blends assured higher tensile strength, thermal ageing resistance, ozone, tack, green strength, air permeability, tear strength etc. compared to pure IIR compound. The compatibility studies on IIR/EPDM blends in different ratios were investigated by viscometric methods [66]. The improvements in dynamic mechanical properties in blends were expected in case of blending compatible rubbers at appropriate ratios [66].

CIIR has many attributes of butyl polymer unit because of its saturated backbone structure of poly-isobutylene. The chlorine functionality in CIIR makes it more reactive to vulcanization system compared with butyl rubber. The CIIR vulcunizates have excellent resistance to weathering, high permeability to chemicals and gases, improved chemical resistance, enhanced thermal stability, good processing properties, better cure versatility and compatibility with other rubbers as well as sealing properties. The chlorination or halogenations of butyl rubber to form reactive allylic halide functionality enhanced the rate of curing and compatibility with other elastomers. Additionally, the presence of polar group in CIIR enhances the adhesion property compared to that of butyl rubbers.

Cong Li *et al.* [67] investigated on dynamic mechanical and thermal analysis of Chlorobutyl rubber blends. It was found that broadening of tanδ peak occurred over a broad range of temperature and two kinds of relaxations occurred in the blends. The formation of intermolecular hydrogen bonding seen from FTIR spectrum revealed strong interaction in the blends.

Jinrong Wu *et al.* [68] investigated the molecular mobility through glass transition range of CIIR by dynamic mechanical analysis and dielectric spectroscopy studies. An asymmetrical broad structure of loss tangent peak indicated two relaxation mechanisms with a maximum peak at higher temperature and shoulder peak at lower temperature region. It was revealed that shoulder peak at lower temperature regions represents the α-process originating from local segmental motion. The maximum peak represents a slow process arising from the mobility of longer chain segments. The results from dielectric studies also showed shoulder peak similar to that of DMA analysis. Positron annihilation lifetime spectroscopy (PALS) was also used to study free volume content and microscopic characteristics. It was pointed out that effective chain packing in CIIR, increased fragility and broadened transition region. During the thermal process, the thermal motion of longer chain segments weakens the interaction of molecular chains leading to faster expansion of free-volume holes after the slow process. The effective chain packing density and slow expansion of free-volume holes inevitably retards the segmental mobility of long chain CIIR molecules. The movement of long chain and local segments of CIIR separates from each other in time or temperature scale. This phenomenon may be the prime reason for the two-peak structure besides low intermolecular interactions.

Traian *et al.* [21] reported about the macroscopic changers on structural characteristics in butyl and halogenated butyl rubber when subjected to gamma radiation upto 0.5 MGy. The irradiated structure showed a complex chemistry comprising simultaneous crosslinking and/or chain scission of macromolecular chains and process involving free radicals. The evaluations of structural variation were carried out by oxidation level, gel content, unsaturation and halogen distribution dependent on radiation dose. The physical properties, viz., mechanical, electrical, optical, etc. and the chemical structure of IIR and CIIR vulcunizates were profoundly altered by high energy ionizing radiations either in inert or oxygen environments. The sudden decline in relaxation behavior of CIIR after irradiation was attributed to radiation-induced degradation effect.

The same research group [69] has evaluated compatibility of EPDM/butyl rubber blends after gamma irradiation. The oxidation stability was characterized using gel content measurements. The IIR content are more susceptible to produce more free radicals than EPDM due to its low radiation capacity and it was found that free radicals produced was grafted on the macromolecules of EPDM or they would recombine to form initial structure leading to less variation in weight fraction after irradiation. The saturated structure of the blend also facilitated a barrier against oxygenative degradation. This study also described about enhanced chemical stabilization by cross linking up to 200kGy and stability of polymers against oxygen uptake and oxygen consumption derivative with time.

M Madani [70] investigated on the influence of gamma irradiation on electrical, swelling and mechanical properties of cross linked IIR, EPDM and their blends filled with carbon black. Carbon based fillers like carbon black, metallic particles and carbon fibers generally improve physico-mechanical properties apart from enhancing conductivity. It was portrayed that changes caused by the degradation of irradiated polymers are variations in molecular weight distributions and production of volatile products. It was noticed that elastomer chains were kept in phase with the applied sinusoidal stress field because of decline in relaxation of its segmental mobility as a result of radiation induced crosslinking. The increase in tensile modulus and decrease in elongation at break was attributed to cross-linking. The average molecular weight and crosslink densities showed a reverse U-shaped behavior in relation to benzene solvent uptake as a function of irradiation dose. Studies on ozone resistance, air permeability and thermal ageing of butyl rubber and EPDM/IIR blends were carried out by Saha deuri *et al.* [71] to evaluate its service performance. It was revealed that oxidative degradation is reduced to an extent in EPDM/IIR blends when compared with IIR due to the protective nature of EPDM against ageing conditions. The ozone resistances apart from thermal, air retention and mechanical properties were enhanced with EPDM content and the optimum properties were found at 85:15 blend ratio (IIR:EPDM). By blending IIR with EPDM, the drawbacks of heat softening at service temperature 120°C were eliminated. The same research group has carried out thermal and thermo-oxidative degradation of IIR/EPDM blends of different compositions from 60°C to 700°C. It was concluded that absorption of oxygen by EPDM led to formation of comparatively stable peroxides resulting in reduction of oxidative degradation in the structure of IIR/EPDM blend [72].

19

The behavior of polymer blends towards high energy radiations depends on how the constituents of the blend interact with radiations [73]. The effect of radiation interaction on an elastomer results in linear energy transfer which generates excited molecules that are homiletically split to provide free radicals. These radicals are further involved in subsequent reactions discussed in previous sections [69] [74]. The exposure of elastomer blends to thermal or radiation ageing brings diverse configuration of bond distribution along the molecular backbones [75]. There are very few literatures reported on compatibility and properties of EPDM-CIIR rubber blends. The summary of various studies on EPDM-CIIR blends is presented in Table 1.1 below.

1.8. Review of EPDM/CIIR blends

Table 1.1 List of research works based on EPDM/CIIR blends

Properties studied	Inferences	References
Effect of EPDM grade on mechanical properties and ageing resistances	The EPDM rubbers 301T and NDR 4640 grade showed additive behavior and synergistic effect respectively. The cure rate for EPDM-CIIR blends with higher compositions of EPDM is more due to presence of ENB (un saturation due to diene content). Synergism of two components in the blend, interphase crosslinking and compatibility was the reason for enhanced mechanical properties.	Sunil jose and Rani Joseph [76]
Influence of precurring of CIIR on mechanical properties of EPDM/CIIR blends	The optimum pre curing was determined on the basis of variation in mechanical properties and pre cured CIIR substantially improved mechanical properties of EPDM-CIIR blends. The cure migration between the phases was reduced, thus improving co-curing apart from blend properties. In the blend containing pre cured EPDM, over curing takes place in EPDM phase leading to poor	Sunil jose, Anoop anand and Rani joseph [77]

20

	interfacial linking and inferior properties. Blends obtained by partial curing of CIIR at 20% of the cure time at 170°C possessed better properties.	
EPDM/CIIR blends cured with reactive phenolic resin	The blends prepared by phenolic resin curing method exhibited superior mechanical and hot air ageing resistances and have potential applications in high-temperature engineering products like conveyer belts, curing bladders etc. By resin vulcanization, crosslinking of both phases and at interphases were improved remarkably. Fractured surface SEM images exhibited co-continuous morphology which substantiates higher ageing and thermal degradation resistance.	Sunil jose, Anoop anand and Rani joseph [78]
Rheology, air permeability, thermal diffusivity and stability of EPDM/CIIR blends	The degree of crosslinking with different types of curing agents and blend compositions on rheological and processability were measured. Blends with 50:50 ratio (EPDM:CIIR) exhibited steep decline in shear viscosity with increasing shear rate. The partially pre cured CIIR content in the blend exhibited highest shear viscosity. Slightly higher values shown for resin cured blends compared with sulfur cured counterparts can be attributed to the entangled long chain resin molecules. It was also elucidated that temperature has negligible effect on shear viscosity i.e. for controlling flow properties of EPDM/CIIR blends. Air permeability decreased as a function of CIIR content because of the inherent compact structure of CIIR. The blends containing EPDM and higher percentage of EPDM exhibited	Paul and joseph [79]

	higher thermal diffusivity and stability which was due to the increased interaction between EPDM and carbon black. The superior thermal stability of resin cured EPDM/CIIR blends reinforced with 1 phr of boron nitride was due to the formation of C-O-C bonds, which are more thermally stable than C-S-C bonds in sulfur cured blends.	

To obtain good properties in blends, it is necessary that EPDM/CIIR blends are compatible [57][80] and have minimum interfacial tension between the two polymer phases. The solubility parameters of EPDM and CIIR are similar making them miscible blends without the aid of compatibilizer.

Further improvement in properties of rubbers can be achieved by reinforcing with nanometric fillers [81] [82]. The nanoscale fillers which are considered to be important for reinforcing elastomers are layered silicates or nanoclay (Montmorillonite [83]), nanotubes (carbon nanotubes [84]), nanosilica [85], carbon nanomaterials (graphene) [86], polyoligometricsilsesquioxane (POSS) and so on.

1.9. Elastomer nanocomposites

Elastomer nanocomposites represent a class of materials that have assumed great importance in recent decades and are in the focus of extensive research [87]. The advent of nanomaterials has created awareness towards research and development of elastomer nanocomposites possessing novel material properties. The nano materials have at least one dimension in the nanosize range (1 to 100 nm). They own properties distinct from bulk materials due to its nanoscale atomic dimension, larger specific surface area and nonexistence of crystalline boundary conditions. In nanocomposites, the significant improvement in properties is attained at very low nanometric filler content (less than 5wt %) [88], [89]. These improvements in properties include mechanical, solvent barrier, flame retardancy, thermal stability, decrease in permeability, etc. Apart from the properties of filler and matrix, dispersion, orientation of nanofillers as well as larger interfacial

22

area provided by nanoscale fillers play a vital role in determining the property of elastomer nanocomposites [90]–[92]. Additionally, the size, shape, type, chemical composition, effective volume fraction and specific surface area of nanoscale fillers have significant role in the degree of property enhancement [87]. The number of particles per gram in nanomaterials is very high compared to that of micro fillers. The augmentation in filler-rubber interfacial interaction in nanocomposites at a significantly lower level of filler loadings compared to conventional reinforcing fillers used in rubber industry can be attributed to superior specific surface area of the nanofiller. Many researchers have delineated the fact that improved interfacial interactions between nanofiller and matrix are the reason behind significant enhancement in mechanical, thermal stability, barrier and viscoelastic properties of elastomer nanocomposites [52], [82].

Conventionally, elastomers reinforced with nanometric fillers like nanoclay, nanosilica, carbon nanotubes, carbon nanofibres, graphene, metal oxides, etc. are used in many engineering entities [93]. For past several decades, scientists and researchers are paying attention on the novel properties of elastomer nanocomposites that have prominent role in wide range of applications in aerospace, nuclear, radiation shields, automotive, high temperature conveyer belts, launch vehicles, civil engineering, biomedical applications, to name a few. Elastomers filled with nanofillers is a rapidly growing area of nano-engineered materials as it offers light weight alternatives and diverse properties with value addition compared to conventional filled elastomers. The first instance of the application of nanofillers was when Toyota used "nylon-nanoclay" in their automobiles in 1990s [94]. The nanocomposites of nylon developed by them exhibited superior mechanical characteristics, improved barrier properties and higher heat distortion temperature than unfilled nylon [95].

The advancement of nanomaterials, synthesis and characterization have shifted the focus towards developing advanced elastomer nanocomposite possessing properties superior to conventional micro-scale composites like efficient reinforcement without the loss of ductility, increase in modulus by factor of 3, higher impact strength, resilience, damping properties, abrasion resistance, barrier, air permeability, enhancement in heat deflection temperature by 100-170°C and so on. Though conventional fillers like silica and carbon black impart good reinforcement on rubbers [96], certain drawbacks at high filler contents like increase in weight at higher filler content (greater than 40 wt. %) and reduced processability led to the requirement of rubber nanocomposites. The improvement in properties in elastomer nanocomposites can be attributed to the following factors [97]:

1. Extensive interfacial area per volume of particles (10^3-10^4 m^2/mL)
2. Low percolation threshold (~0.1-2 vol.%)
3. Large number density of particles per unit volume (10^6-10^8 particles/μm^3) and
4. Short distance between the particles (10-50 nm at ~1-8 vol. %) [95]

The reinforcements imposed by nanofiller in elastomer chain conformations will alter the molecular mobility, relaxation aspects and thermal stability, apart from property enhancements. The state of distribution and dispersion of nanofiller throughout elastomer matrix is one of the key parameters that decide the efficiency of reinforcement in terms of properties. The structural configuration at the molecular level and paves the way for maximizing interfacial region leading to efficient stress transfer between filler and matrix seen in nanocomposites and thus projects the potential of defining new material properties [98]. The major challenge in rubber nanocomposites is to attain homogenous dispersion and complete exfoliation of nanofillers in the elastomer matrix because rubber matrix is not melted during processing unlike thermoplastics. There are many scientific literatures available in this field based on elastomers [99], thermoplastics [100], thermosets [101], high performance polymers [93], etc.

1.10. Types of nanofillers

The basic classification of nanocomposites [81], [84], [97] is based on its dimentionality viz. isodimentional nanoparticles having all the three dimensions in nanorange (ex: nanosilica, nanotitanium, nanocalcium carbonate, etc.), two dimensional nanofillers like nanowhiskers, carbon nano fibers, nano tubes, etc. that have two-dimensions in the nano range and in an elastomer matrix form a one dimensional nanocomposite, one dimensional nanofillers having individual nanoscopic layers in one dimension and form a multi-dimensional composite when incorporated on an elastomer matrix (ex. layered silicates or nanoclay). The types of nanofillers based on geometric aspects are shown in Figure 1.6 below.

24

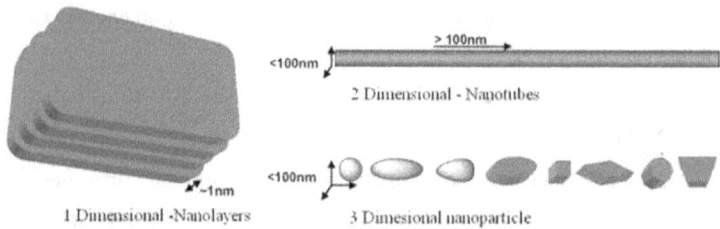

Figure 1.6. Classification of nanofillers based on geometry [81]

1.10.1. Nanofillers with platelet like structure (nanoclay or layered silicates)

Layered silicate or nanoclay is commonly used nanomaterial for preparation of rubber nanocomposites that belong to phyllosilicate family. Compared to other nanomaterials, layered silicates are easily available at a cheaper price. The layered silicates or nanoclay have higher aspect ratio of the order of 100-1000 due to its sheet like platelet structure. The polymer/silicate nanocomposites have attractive nanofiller for mechanical, barrier and other properties of polymers because of their high aspect ratio and larger specific surface area. Their unique layered 2D structural arrangement consists of a two-dimentional layer of two fused silicate tetrahedral plates with an edge shared octahedral sheet of a metal atom such as Al or Mg. Montmorillonite (MMT) crystal lattice consists of 2 D layers with a central octahedral sheet of magnesia or alumina fused to two external silica tetrahedrons by the tip, so that the oxygen ions of the octahedral sheet also belong to the tetrahedral sheets. The thickness of the sheet is around 1nm and length of the sheet may vary from 100nm to several microns depending on specific layered silicate. These platelets or layered silicates organize themselves into stacks leading to a regular Vander Waals gap in between layers known as interlayer space. Layered silicates are characterized by cation exchange capacity (CEC) expressed as mequiv/100 g due to the surface charge. The montmorillonite are the type of clay that possess greater significance in the manufacture of polymer nanocomposites because of its higher CEC and swelling capabilities. The organomodified clay (or organoclay) has enhanced "wetting with the polymer matrix", and has better adhesion with organic polymers and lower surface energy than unmodified nanoclay. Apart from facilitating larger interlayer spacing which provides greater intercalation of rubber chains because of long organic chains, organo modification imparts "functional groups" that can interact more with the rubber matrix.

The dispersion and distribution of layered silicates in the polymer matrix is the key factor behind enhancement in its properties apart from other properties of filler like aspect ratio, specific surface area, geometry and volume fraction. The homogeneous dispersion of layered silicates in rubber leads to better filler-rubber interactions. It also contributes towards improvement in physico-mechanical properties due to well-dispersed morphology of the nanocomposites [102].

The dispersion of layered silicate in the rubber matrix is characterized at three stages, viz., nano, micro and macroscopic scale. The dispersed nanoclay in the rubber matrix possesses three structures, namely, (i) intercalated (ii) intercalated and flocculated and (iii) exfoliated, which are shown in Figure 1.8. In the intercalated structures, clay particles are dispersed in an ordered lamellar structure with larger gallery height via insertion of rubber chains into the gallery. Each silicate layer is delaminated and dispersed in a continuous elastomer matrix in exfoliated structures [103]. X-ray diffraction studies (XRD) and electron microscopic evaluations are common tools for evaluating the morphology of nanocomposites [103], [104]. The nanoclay and rubber remain as distinct phases when rubber chains are unable to intercalate in between silicate layers. The properties of such nanocomposites are very much similar to conventional microcomposite. The nanocomposite has the following divergent characteristics compared to microcomposite:

1. Delaminated layers of nanoclay allows the rubber chains to get intercalated
2. Total delamination and distribution of silicate layers results in exfoliated structures
3. Ionic interactions of clay (edge to edge or edge to face) may lead to flocculated structure

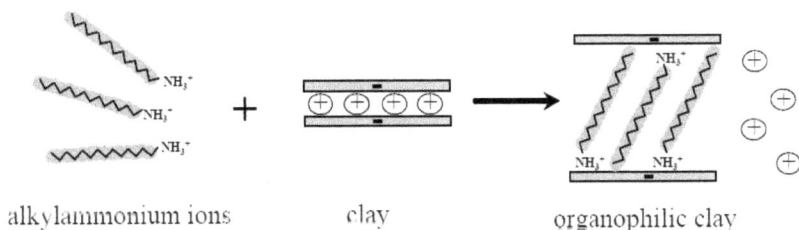

alkylammonium ions   clay   organophilic clay

Figure 1.7. Organomodification of nanoclay [105]

26

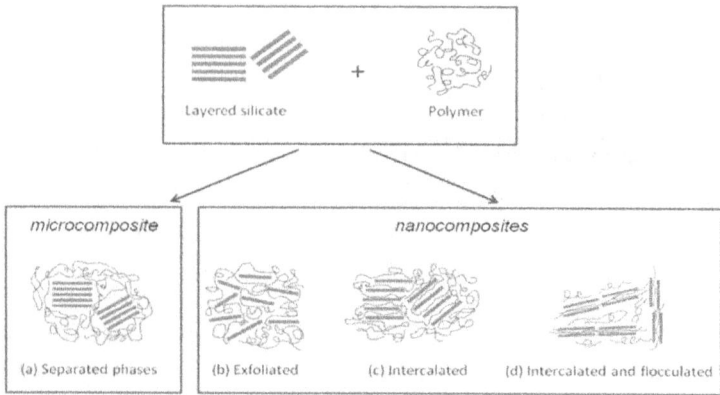

Figure 1.8. Structure of microcomposite and nanocomposite [106]

1.10.2. Particulate or isodimensional nanofiller (nanosilica)

Nanosilica or nanosilicon dioxide particles having nanoparticle size (11-13 nm) and large specific surface area (200 m²/g) are used as reinforcing agents in polymer industry. It has high melting point of 1710°C and boiling point of 2230°C with a low density of 2.19 g/cm³. The chemical structure of nanosilica particles is shown in Figure 1.10. Nano-silicon dioxide particle (nano-SiO_2) is an important inorganic chemical product. It is an excellent raw material for the production of rubber, plastic, paint, printing ink, paper, pesticide and toothpaste. In recent years, remarkable results have been achieved in the study of modifying and toughening polymers and hybrids with inorganic silicon dioxide nano particles [107]–[109].

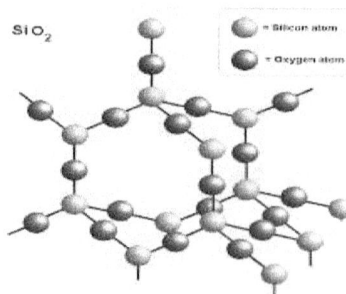

Figure 1.9 Chemical structure of nanosilica particles [110]

27

The covalent bonds between silica and oxygen atoms as seen in Figure 1.10 attribute towards the properties of nano-SiO$_2$. Nano-SiO$_2$ particles have unique advantages of surface interface effect, quantum size effects and macroscopic quantum tunneling effect, special optical and electrical properties and high magnetic phenomenon. Even at high temperature, it has high strength, toughness, and good stability. Nanosilica particles can effectively improve the mechanical properties and heat resistance of the resin matrix. Several scientific literatures are available on the discrete chemical and physical structure of nano-SiO$_2$ particles and their utility in engineering applications. The synthesis of nano-SiO$_2$ is carried out by flame synthesis, reverses micro emulsion and widely used sol gel process. The surfactant molecules are dissolved in organic solvents to form spherical miscells in reverse micro emulsion process. The major negative aspect of this synthesis method is the difficulty in extracting surfactant molecules from the final product and higher cost. The high temperature flame decomposition method or chemical vapor condensation (CVC) is also used for the production of nano-SiO$_2$. The reaction between silicon tetrachloride (SiCl$_4$) and hydrogen and oxygen occurs in this process. Evaluation of particle size, phase decomposition and morphology of nano particles produced by this process are challenging. The hydrolysis and condensation of metal alkoxides (Si(OR)$_4$) such as tetraethyl orthosilicate (Si(OC$_2$H$_5$)$_4$) or inorganic salts (Na$_2$SiO$_3$) in the presence of acidic (HCl) or alkaline (NH$_3$) medium as catalyst occurs in sol-gel method [107-109].

There are hydroxyl groups (–OH) existing on the surface of nano-SiO$_2$, leading to the strong hydrophilicity, high surface activity, silica-silica interaction and poor stability. Therefore, it is difficult for them to infiltrate and disperse in organic medium. Nano-SiO$_2$ particles are prone to agglomerate, forming a chain structure. The chain structure interacts with each other by hydrogen bond to form a three-dimensional network structure in which the intermolecular forces are very strong. The surface treatments with various modifiers like silane coupling agents have been carried out to enhance filler dispersion and intensity of rubber/nanosilica interfacial interactions. Nanosilica particles after surface modification, impart the highest degree of reinforcement amongst all of non-black particulate fillers because of their particle size and aggregated complex structure. Coupling agent can combine inorganic fillers and organic polymer materials by typical physical and/or chemical reactions, which improve dispersion and reduce the surface energy of nanofiller thus enhancing the comprehensive properties of the nanosilica reinforced rubbers [85], [111]. According to the chemical structure, coupling agents can be divided into silane coupling agent, titanate coupling agent, zirconium coupling agent, aluminate coupling agent, bimetallic coupling agents (aluminum zirconium ester, aluminum and titanium composite coupling agent), rare earth coupling agent, phosphorus coupling agent and boron containing coupling agents. Amongst these, silane coupling agents have attained more attention because of their low cost, ease of preparation and use as well as their environmental friendly characteristics.

Hegazi et al. [112] studied the effect of silane modified nanosilica on acrylonitrile butadiene rubber in resisting gamma radiation. Quang et al. [110] reported on thermo-mechanical properties of NR/NBR rubber nanocomposites reinforced with silane modified nanosilica and found increase in thermal stability due to presence of nanosilica in the matrix. Stevenson et al. evaluated the effect of irradiation on nanosilica filled PDMS elastomer [113]. Interfaces between PDMS and nanosilica played a vital role in resisting irradiation ageing effects. Recently, Zhang et al. [114] reported about the utility of silica nanoparticles in modified epoxy resins for attaining stability against high gamma-radiation. Vojislav et al. [115] studied the effect of gamma irradiation on thermo oxidative behavior of nanosilica based polymer composites. It was reported that incorporation of thermally stable nanosilica raised the temperature of thermo-oxidative degradation. The possible mechanism for enhancement in resistance towards degradation was attributed to the fact that silica nanoparticles inactived terminal –OH groups which participate in degradation process [116]. The attachment of an organic moiety with nanosilica surface, evident from the formation of Si-O-Si and Si-O-C bonds from FTIR spectroscopy after irradiation exhibits material modification in the polymer matrix.

1.10.3. Carbon nanotubes (CNT)

The carbon nanotubes (CNTs) are employed as reinforcing nanofiller in many polymers due to their unique characteristics compared to other reinforcements, since their innovation in 1991 by Iijima [117], [118]. CNTs belonging to fullerene family are needle-shaped tubular single crystals cylindrical in shape made of sp^2 carbon atoms from graphene sheets [84], [119]. The properties of the CNTs depend on atomic arrangement, chirality and aspect ratio of the tube and overall morphology.

CNTs can be either single (SWCNT) or multi walled (MWCNT) as represented in Figure 1.11. CNTs are considered as major reinforcements in polymer composites due to their remarkable mechanical, thermal and electrical properties. The remarkable mechanical properties of CNTs (~100 times greater than steel at a fraction of mass) were revealed from experimental and theoretical studies [93], [120]. SWCNTs and MWCNTs have lower densities than steel and aluminum. CNTs also possess good resilience that enables them to bend at large angle without much deformation [121], [122]. Additionally, they show high young's modulus and electrical as well as thermal conductivities comparable to diamond. They also exhibit good thermal stability and negligible thermal expansion coefficient.

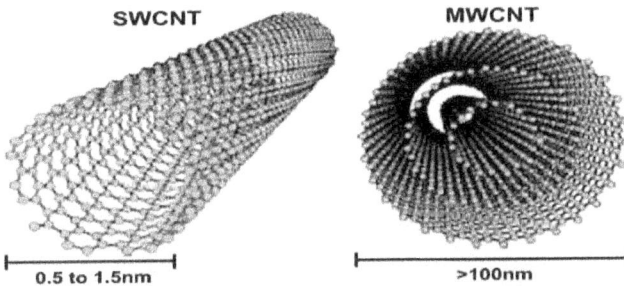

SWCNT   **MWCNT**

0.5 to 1.5nm   >100nm

Figure 1.10 Single and multi walled CNTs [118]

SWCNTs are 2-D graphene sheet having hexagonal array of carbon atoms rolled into a tube. MWCNTs contain multiple layers (2 to ≤ 30) of concentric graphene cylinders. The ends of the tube are capped by fullerene-like structures. The CNTs can have a variety of structures like chiral, armchair or zigzag, as shown in Figure 1.12, based on the twist along the length of the rolled graphene [84], [119].

CNTs conduct heat at 3000-3500 W/mK, which is very high compared to polymer. Polymers are generally insulators with low thermal conductivity of around 0.1 W/mK. The length to diameter ratio of 28,000,000:1 is larger than that of any other reinforcing material [93], [117]. The discrete properties of CNTs make them useful in many applications like electronics, nanotechnology, optics and other fields of materials technology as well as civil and agricultural applications.

30

Though they impart extraordinary mechanical properties, thermal conductivity, electrical characteristics and unique structural characteristics, their final usage in any product applications in industrial as well as academic research may be limited by their potential toxicity and variations in properties when subjected to chemical and acidic ageing environments [118], [123].

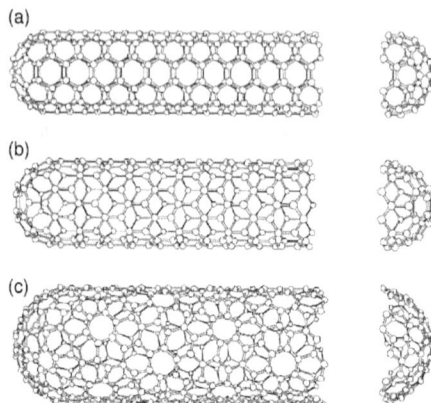

Figure 1.11 Schematic models for (a) arm chair (b) zigzag and (c) chiral arrangement of SWCNTs [95]

The macroscopic properties of polymer – CNT composites are determined by the distribution and alignment of CNT in the polymer matrix. CNTs pose challenges in dispersion because they have strong tendency to form agglomerates resulting in poor dispersion in the matrix [118]. The reason for this can be attributed to the fact that van der Waals forces are strong in CNTs [124]. In addition, energy input to disperse the CNTs tends to break them into smaller segments decreasing their aspect ratio in the final composite with simultaneous enhancement in their dispersability. The nanotubes are chemically treated (surface modification or functionalization) to provide covalent and non-covalent functional group attachments to the structure of CNTs [93]. Functionalized CNTs have better dispersion in organic solvents and polymers as well as enhance the strength of the CNT-polymer matrix interfacial interactions.

Several scientific literature have reported that reinforcement of polymer by nanofillers also improved the radiation resistance of the nanocomposites [15]. Kobra *et al.* [125] evaluated the effect of nanoclay on thermal resistance, tensile properties and gamma radiation shielding performance of polymer composites. Gamma attenuation tests were evaluated by ^{192}Ir, ^{137}Cs and

31

[60]Co sources. Nanoclay has replaced the drawback of lead monoxide in radiation shielding applications. The best radiation resistance properties were obtained for nanocomposite with 5 phr nanoclay. XRD pattern showed no diffraction peak for the nanocomposite revealing complete exfoliation of Cloisite 30B nanoclay. Thermal decomposition temperature was higher for nanocomposite. The effect of clay in epoxy based ternary nanocomposites for radiation shielding application was also reported by some researchers [126], [127]. It was revealed that rapid expansion of nuclear technology necessitated the introduction of new light weight and cost efficient polymer composites [121], [128]. In this context, polymer nanocomposites, owing to their prominent physico-mechanical properties, ease of processability, light weight, etc. have attracted attention for radiation sheilding [113], [129].

1.11. Review on EPDM nanocomposites

EPDM rubber (M-class) nanocomposites [104] have attracted considerable interest in the last two decades due to their properties. Several researchers have reported improvements in mechanical, thermal, barrier and degradation-resistant properties in polymers reinforced with layered silicates (nanoclay) [15]. Seyed *et al.* [104] studied the effect of exfoliated organomodified clay on the structure and mechanical properties of EPDM rubber. EPDM/organomodified clay composites prepared by melt blending method were characterized by XRD and TEM analyses. The increase in storage modulus and decrease in tanδ peak from dynamic mechanical analysis can be attributed to maximum interfacial adhesion between EPDM and nanometric layered silicate surface which restrict the segmental mobility at the interfaces. From static mechanical tests, it was noticed that tensile strength and modulus were increased after reinforcement. It was also revealed that thermal stability was also enhanced as a function of nanoclay content. The presence of nanoclay acted as barrier to permeate volatile degradation products from matrix [130].

Chang *et al.* [103] reported the properties of EPDM/organoMMT hybrid nanocomposites. The rubber molecules were found to be "intercalated into the galleries of organoMMT and the silicate layers" were uniformly dispersed as platelets of 50-80 nm thickness in EPDM matrix. A stronger filler-rubber interaction was evident from DMA which was manifested by decline in loss tangent peak. Apart from reinforcing, well exfoliated nanoclay layers also act as a physical barrier against crack propagation, leading to increase in tear as well as gas permeation resistance.

Mascoud *et al.* [131] studied organoclay composites of polypropylene/EPDM blends and found that the layered silicates improved gas barrier properties by decreasing area available for diffusion and providing tortuous path for solute permeation through the polymer. Karinbir *et al.* [132] evaluated the effect of nanoclay and compatiblized EPDM rubber on impact strength of epoxy glass fiber composites. The significantly enhanced impact behavior of nanocomposites was attributed to multi-scale reinforcements provided by nanoclay and elastomer as well as stress transfer capability of silicate layers. He *et al.* [133] reported exfoliation of organic modified montmorillonite (oMMT) in EPDM enhanced tensile strength (TS), elongation at break, ablative properties and flame retardancy while maintaining the structural stability for solid rocket motors insulation applications. Effects of nanoclay reinforcement on EPDM rubber for evaluating its sealing characteristics (decompression failure analysis) were explored by Elaheh [134]. The result revealed that incorporation of nanoclay at some compositions could slightly reduce the strength of rubber. However, more stable modulus at varying strains was obtained along with least change in hardness as well as reduction in permeability. The performance of rubber with nanoclay under compression was enhanced, which is essential in seal applications.

EPDM/organoclay nanocomposites were also studied by Peiyao *et al.* [135] and it was reported that tensile strength increased up to 12.3 MPa for EPDM with 3 phr nanoclay which was five-fold higher than EPDM and higher than conventional composites with 1.5 phr CB. Melt extrusion lead to better exfoliation of modified clay mineral particles in EPDM. There are several methods of preparing elastomer nanocomposites like solution mixing, latex blending, internal mixing and open mill mixing [136]. Solid state mixing using internal mixers promises a direct and environmental friendly technique for preparing elastomer nanocomposites [136]. Hermann *et al.* [137] described two-step mixing in an internal mixer followed by the addition of curative on a two-roll mixing mill for developing elastomer nanocomposites. Gatos *et al.* [138] have studied on the effect of processing parameters (mixing type and temperature) and formulations on the properties of EPDM nanocomposites. It was noticed that "open two roll mixing resulted in poor dispersion of nanoclay in the rubber compared to compounding in an internal mixer".

Zhang *et al.* [139] studied the enhanced mechanical properties, thermal stability and smoke suppression effect of dendrimer modified montmorillonite (DOMt) on EPDM. Baolei *et al.* [140]

reported a novel route to develop EPDM/MMT nanocomposites by modifying MMT using maleic anhydride (MA) which is an intercalation agent for MMT and aids ease of curing for EPDM as well as act as compatibilizer for both EPDM and MMT. The structural, thermal and material properties were the three aspects investigated based on the effect of MA modified MMT in nanocomposites. Nanocomposites exhibited better tensile strength, elongation, modulus, dynamic properties and thermal degradation behavior.

There are only few studies that laid systematic investigations on incorporation of nanofillers in EPDM for gamma radiation resistance applications [141] [105]. The comparative study of the effect of gamma irradiation on EPDM/clay nanocomposites and EPDM/pristine clay composites was carried out by Seyed et al. [105]. The dynamic mechanical thermal analysis (DMTA) demonstrated shift in α-relaxation peak and storage modulus towards higher temperature as a function of irradiation dose. The exposure of EPDM hybrids to gamma rays enhanced tensile strength of samples at low radiation dose (up to 200kGy) due to the dominance of crosslinking effect and then marked decrease in tensile strength as the dose rose up to 100kGy because of the dominance of chain scission. However, at high irradiation doses, the reduction in tensile strength for the nanocomposites took place at a slower pace. The tensile strength remained constant at the irradiation dose of more than 100kGy. In case of unfilled and conventional EPDM composites, tensile strength reduces progressively at more than 100kGy dose. EPDM nanocomposites exhibited superior irradiation resistance properties than unfilled and EPDM composites. It was concluded that well dispersed nanoclay and its interaction with the rubber matrix protects against radiation induced degradation.

Like organomodified layered silicates, nanosilica particles are also used as reinforcing fillers in rubber for achieving enhancement in properties like mechanical [142], viscoelastic, thermal [143], flame retardancy, solvent barrier properties, radiation ageing resistances etc. The interfacial interaction and dispersion of nanosilica particles in the elastomeric matrix is the key to explore with the advantage of nano reinforcement. Numerous scientific literatures are available which deals with incorporation of conventional fillers like silica in rubbers.

Nano-SiO$_2$ possess excellent structural characteristics based on its chemical nature [144] as discussed in previous section. The primary reason for enhancement in mechanical properties, thermal stability, rheological characteristics to name a few in polymer-nanosilica composites is due to the presence of functional groups present on its surface [145]. The surface hydroxyl groups of silica including vicinal, isolated and geminal silanols are believed to play a critical role in aforementioned property improvement [146]. These new class of materials namely organic-inorganic hybrids, afford to combine both the advantages of organic material as light weight, good moldability and flexibility, and of inorganic fillers such as high heat stability, strength and chemical resistance [147][108].

Surface treatments with various modifiers like silane coupling agents as discussed in the previous sections have been carried out to enhance filler dispersion and elastomer/nanosilica interfacial interactions [148]. Reinforcement with nanosilica significantly improved dynamic characteristics as well as mechanical, barrier, thermal and impact properties in Ethylene Propylene Rubbers (EPR) [85]. The surface modification of nanosilica also enhanced the non-linear viscoelastic behavior (Payne effect) of nanosilica filled rubber by decreasing the de-bonding of polymer chains from filler interface [149].

Only limited literature are available on gamma irradiation effect of nanosilica filled EPDM rubbers. Madani [150] reported the thermal analysis and electrical properties of EPDM composites incorporated with micro and nanosilica. Remarkable heat resistance and mechanical property was observed for nano and micro composites compared to pristine EPDM rubber. The distribution of nanosilica particles in the elastomeric matrix inhibited thermal degradation of vulcunizates which led to enhancement in thermal stability. Suzana et al. [151] reported rheometric characteristics, cure kinetics, morphology and mechanical properties before and after thermal ageing of nanocomposites of EPDM/NBR blends. The specific chemical interactions between rubber and filler were characterized using FTIR spectroscopy. The particle surface of silica possess hydrophilic silanol or hydroxyl groups which results in the formation of strong filler-filler interactions via hydrogen bonding. The formation of bond between acidic surface of silica particles and basic nature of nitrile group (-CN) in nitrile rubber resulted in strong filler/rubber interactions. The increase in intensity of band corresponding to $-CH_2$ deformation from FTIR spectra was due to strong interaction between the filler and matrix. The higher accessible sites provided by nanosilica particles in the polymer matrix takes high stresses applied to the rubber matrix [152]. Additionally, the presence of several functional groups on the surface treated nanosilica particles would support physico-chemical interactions at the filler/rubber interfaces [153] [154]. Thus, nanosilica particles are promising candidate for several heat shield [155] and gamma radiation resistance applications [151] [156].

Several scientific literature are available on the influence of nanosilica in resisting gamma radiations in elastomeric [157][158][159][156] and other systems [160][161]. However, there are no literatures are reported on nanosilica reinforced EPDM rubbers for retarding effects of gamma irradiation.

1.12. Review on CIIR nanocomposites in view of barrier properties

The researchers developed nanocomposites of CIIR for applications involving barrier resistance after the exploration of nanofillers and its functional benefits on elastomer matrix was expounded. Some of the research works carried out on CIIR nanocomposites are tabulated in Table 1.2.

Table 1.2. Research works carried out on CIIR nanocomposites

Properties studied	Inferences	References
Immobilizing polymer chains in CIIR nanocomposites	The organic modification, volume fraction and lattice spacing of added nanoclay have a profound effect on the constrained polymer volume in CIIR nanocomposites. Studied on different types of nanoclay (Cloisite 10A, 15A and 20A). The enhanced volume of constrained regions obtained from DMA led to increase in glass transition temperature and reduction in loss tangent peak	Saritha *et al.* [162] [92]
Effective gas and VOC barrier property and modeling approaches.	CIIR nanocomposites were prepared by organomodified Cloisite 15A and characterized using XRD and TEM. Permeation properties were modelled using Nielson, Cussler and Guslev and Lusti models.	

	Cussler model showed satisfactory agreement with lower experimental tortousity factors.	
Rheological behavior of clay incorporated natural rubber (NR) and CIIR.	Rubbers reinforced with Organoclay (Nanomer I.44P) were investigated using linear dynamic viscoelastic measurements. The WAXD and TEM showed an intercalated and agglomerated morphology for CIIR/clay nanocomposites. CIIR possessed higher viscosity, more polarity due to chlorine atoms and steric effects than NR. The inherent resistance of CIIR towards barrier properties is due to its chemical structure (close packed structure due to steric crowding associated with methyl groups).	Ajesh *et al.* [163]
Barrier properties of CIIR nanoclay composites	Solvent barrier against exposure to various chemicals are studied. The sorption experiments on CIIR nanocomposites were carried out in three solvents of different cohesive energy densities (CED) till equilibrium. The decrease in rate of solvent uptake and diffusion coefficients indicated enhancement in barrier properties. The presence of polar group in CIIR renders more barrier against permeation of chemicals.	Sridhar V and Tripathy D [164]
Experiments and modeling of non-linear viscoelastic responses in NR and CIIR nanocomposites	The filler-filler and filler-rubber interactions of nanocomposites of poly-isoprene rubber (NR) and chlorine substituted poly isoprene-isobutyl rubber (CIIR) were studied by Payne effect [165]. The theoretical perspectives were studied by Maier-Goritz and traditional Kraus theory. The solvent diffusion experiments and modeling using Kraus, Lorentz-Parks and Cunneen-Russell equations	Ajesh *et al.* [166]

	were also carried to corroborate transport coefficients with nanofiller interfacial interactions with rubber matrices. Tortous path and decreased free volume by the nanoclay makes it difficult for the solvent to permeate through the CIIR matrix.	
Influence of varieties of organo clay on constrained volume of CIIR nanocomposites	Different clay moieties have been chosen to study the variations in concentration of modifier, modification type and d-spacing. Well dispersed and exfoliated structure of CIIR nanocomposites were evaluated by XRD and TEM. Storage modulus increased with increase in filler content due to the enhancement of stiffness in the nanocomposite. The higher surface to volume ratio of filler enhanced interfacial and restricted segmental mobility. Comparison of dynamic mechanical properties and constrained volume of elastomer chains [167] as a function of filler type and content was also investigated.	Saritha A and Joseph K [168][169]
Effect of solvent interaction parameters in tailoring properties of nanocomposites.	CIIR nanocomposites prepared by solvents with varying CED were studied. Effect of solvent-clay and solvent-rubber interaction parameters in determining properties was investigated. The results were interpreted using thermodynamic concepts and correlation between mechanical and solvent interaction parameters was derived. It was delineated that strong relation between properties and solubility parameters exist in solution mixing process.	

1.13 Justification for the research

In nuclear fuel reprocessing facilities, elastomers find utility in o-rings, gaskets, master slave manipulators, airlock beadings, hatches, etc. These components are subjected to withstand intense gamma radiation and hydrocarbon solvent ageing environments. Among elastomers, EPDM has the highest resistance to gamma radiation and is the material of choice for the former mentioned applications. Though EPDM has very good radiation resistance, it cannot withstand alkyl hydrocarbons that are used in reprocessing of spent nuclear fuels. In this context, blending EPDM rubber with cholorobutyl rubber (CIIR) offers a promising solution to improve the product life of components that are simultaneously exposed to radiation and hydrocarbon aging.

In the very few literatures reported on properties of EPDM–CIIR blends, the focus has been on the effect of processing parameters and compatibilization on static mechanical, thermal, rheological, and air permeability properties. The studies on mechanical, solvent transport, thermal, gamma radiation and hydrocarbon ageing properties of EPDM-CIIR blends are scarce in literature. The present study focuses on developing EPDM/CIIR blends for nuclear fuel reprocessing applications involving simultaneous exposure to both gamma radiation and paraffinic hydrocarbon solvents. The properties of EPDM-CIIR blends can be further improved by incorporating nanosized reinforcing fillers. The use of nanofillers at a very low loading makes it possible to produce fundamentally new elastomeric materials with a range of unusual mechanical and physical properties. Several researches were carried out on reinforcing elastomers with various nanofillers like nanoclay, metal nanoparticles, nanosilica, carbon nanotubes, etc. for numerous applications in aerospace, nuclear, defense, etc. From the literature review of EPDM and CIIR based nanocomposites, it was revealed that there are no scientific literature reported on nanocomposites of EPDM-CIIR blends. The enhancement of mechanical properties, barrier properties and resistance to different cumulative doses of gamma radiation upon incorporation of nanoparticles are yet to be investigated. So far, studies on the viscoelastic properties of EPDM-CIIR nanocomposites have also not been reported.

In the present thesis research, the effect of organomodifed nanoclay reinforcement on EPDM-CIIR blends were evaluated based on the enhancement in static mechanical properties, viscoelastic characteristics, thermal degradation, transport characteristics and effect of cumulative doses of gamma rays from ^{60}Co source. Moreover, this study aims to explicate the reinforcement effect of silane grafted nanosilicon dioxide or nanosilica particles in EPDM-CIIR blends for improvement of the mechanical, viscoelastic, thermal and transport characteristics as well as behavior after exposure to different cumulative γ-radiation.

Furthermore, this study is focused to evaluate the effect of multi-walled carbon nanotubes reinforcement on cure behavior, mechanical properties, transport characteristics and radiation aging of EPDM based blends. The role of nanofillers in retarding radiation ageing effects is also detailed in this thesis research. Preliminary studies on carbon black-nanofiller hybrid composites of EPDM-CIIR blends for product applications are also discussed in this thesis.

1.14 Scope and Objectives of the thesis

The overall objective of this thesis is to develop and evaluate properties of nanocomposites of EPDM/CIIR blends for simultaneous radiation and hydrocarbon solvent environments in nuclear fuel reprocessing facilities. The specific objectives include:

- To optimize the blend ratio of EPDM/CIIR based on mechanical properties, solvent barrier characteristics and gamma radiation effects

- To develop nanoclay (NC) based nanocomposites of EPDM blends and to study the morphology, cure characteristics, static mechanical and viscoelastic properties, nano reinforcement mechanism, thermal degradation behavior, solvent transport characteristics and effect of gamma radiation exposure

- To investigate the influence of bis(3-triethoxysilylpropyl)tetrasulfide (TESPT) grafted nanosilica (NS) particle reinforcement on the mechanical, viscoelastic, thermal and transport characteristics as well as behavior after exposure to different cumulative γ-radiation doses.

- To study the effect of multi-walled carbon nanotubes (MWCNT) reinforcement on cure characteristics, mechanical behavior, transport properties and radiation aging of EPDM blends.

- To evaluate the properties of EPDM/CIIR carbon black-nanofiller hybrid composites for product application.

1.15 Organisation of thesis

The thesis comprises of eight chapters that includes introduction and literature review, materials and methods which gives an overview of materials, preparation methodology and characterization techniques followed by five chapters which discusses the studies on EPDM/CIIR blends, nanocomposites of EPDM/CIIR blends based on nanoclay, nanosilica and MWCNT, hybrid nanocomposites of blend for product application, and finally a chapter on conclusions and future scope.

Chapter 2

Materials and methods

The specifications of the materials used for the preparation of blends and nanocomposites, EPDM, CIIR, nanofillers and other compounding ingredients are detailed in this chapter. The preparation methodology and characterization techniques are also explained in this chapter.

2.1. Materials

The rubbers and nanomaterials used for preparation of nanocomposites based on EPDM-CIIR blends used in this study are described below.

2.1.1. Ethylene propylene diene monomer (EPDM)

EPDM, a M class rubber have saturated chain with 43% ethylene content, 14.2% ethylidenenorbornene (ENB) as the diene content, Mooney viscosity [ML (1+4) at 125°C] of 47.8 and iodine value 30 was procured from JMF Synthetics India Ltd., Mumbai, India.

2.1.2. Chlorobutyl rubber (CIIR)

CIIR, chlorinated isobutylene isoprene rubber has Mooney Viscosity [ML(1+8) 125°C] of 38 and chlorine content of 1.26 % was bought from Lanxess Pvt. Ltd., Chennai, India.

2.2. Compounding ingredients

2.2.1. Vulcanizing agent

The vulcanizing or curing agent used in evaluating properties of nanocomposites of EPDM/CIIR blends is sulfur. The elemental sulfur (purity 99.6 % min) used in this study is of commercial grade.

2.2.2 Fillers

(a) Nanoclay

Nanoclay (Nanomer 1.34nm) used is surface modified that contains 25-30 wt. % methyl dihyroxyethyl hydrogenated tallow ammonium (density $200\text{-}300Kg/m^3$ Size <= 20 micron) was purchased from Sigma Aldrich.

(b) Nanosilica

Nanosilica or silicon oxide nanopowder (SiO_2) having particle size of 60-70nm, specific surface area of 180-220 m^2/g (spherical nano powder form) was procured from Intelligent Materials Pvt Ltd. (Nanoshel), India. The silane coupling agent (Si-69), used for surface functionalization of nanosilica is bis (3-triethoxysilylpropyl) tetrasulfide (TESPT). TESPT, a bifunctional, sulfur containing organosilane agent with the chemical formulae $(OC_2H_5)_3\text{-}Si\text{-}(CH_2)_3\text{-}S_4$ chosen as coupling agent with the chemical structure as shown below was procured from Ponmani Chemicals & Co., Coimbatore, India.

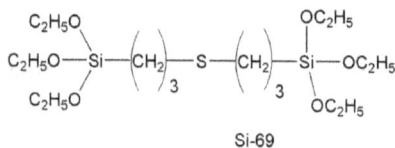

Si-69

bis (3-triethoxysilylpropyl) tetrasulfide (TESPT)

(c) Multi-walled Carbon Nanotubes (MWCNT)

MWCNT modified with 1.8% -COOH was supplied by United NanoTech Innovations Pvt. Ltd, Bangalore, India. The MWCNTs had outer diameter of 20 nm, inner diameter of 16 nm and an average length of 20 μm.

(d) Carbon black

Carbon black used for preparation of hybrid nanocomposites of EPDM/CIIR blends for product application is of High Abrasion Furnace (HAF) grade with particle size of 31 to 200nm.

2.2.3 Activators

The activators used were zinc oxide and stearic acid (commercial grades).

2.2.4 Accelerators

(a) 2-mercaptobenzothiazole (MBT)

The MBT of commercial grade was used as the primary accelerator.

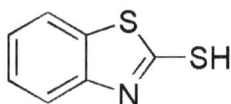

2-mercaptobenzothiazole (MBT)

(b) Tetramethylthiuram disulfide (TMTD)

The TMTD of commercial grade was used as secondary accelerator.

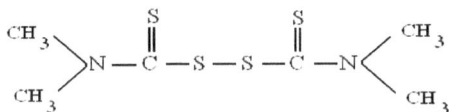

Tetramethylthiuram disulfide (TMTD)

(c) N-cyclohexyl-2-benzothiazole sulphenamide (CBS) and Zinc diethyl dithiocarbamate (ZDC)

Commercial grades of CBS and ZDC with the chemical structures shown below were used as fast accelerators in this study.

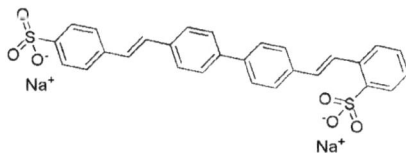

N-cyclohexyl-2-benzothiazole sulphenamide (CBS)

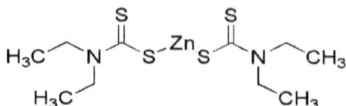

Zinc diethyl dithiocarbamate (ZDC)

2.2.5 Antioxidant

The antioxidant used in this study is commercial grade 2,4-trimethyl-1,2-dihydroquinoline polymer (TMQ) shown below.

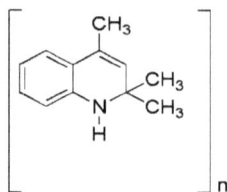

2,4-trimethyl-1,2-dihydroquinoline polymer (TMQ)

2.3. Preparation methodology

A two-stage compounding process was adopted to ensure dispersion of nanofillers in the elastomeric blend. In the preliminary step, masterbatch of rubbers and nanofillers were prepared in a lab scale internal mixer. The second step involved compounding EPDM and CIIR with calculated quantities of master batch and compounding ingredients as per ASTM-D 3182 standards on a laboratory scale two-roll mixing mill.

The masterbatches of EPDM-nanofiller and CIIR-nanofillers were prepared with the weight ratio of 3 parts of rubber to 1 part of nanofiller (3:1 ratio). The EPDM and CIIR were first masticated till the torque was stabilized and then nanofillers were added in the internal mixer. Masterbatches of nanoclay, nanosilica and MWCNTs were prepared separately. The prepared masterbatches were used in calculated quantities to prepare nanocomposites. To ensure cure compatibility pre-

45

cured CIIR compound used in second stage. The CIIR was pre-cured in an air oven for 20% of the optimum cure time of 100% CIIR compound (6 minutes) at 170^0C [77-79].

Table 2.1. Formulation of EPDM/CIIR nanocomposites

Ingredient	phr
EPDM	80
CIIR	20
Zinc Oxide	4.0
Stearic Acid	1.5
Nanofiller (nanoclay, nanosilica or MWCNT)	Varied
Carbon black	50
Naphthenic oil	10
MBT	1
TMTD	0.5
ZDC	0.5
CBS	1
Sulfur	1
Antioxidant	1

In the second stage, the EPDM rubber, precured CIIR and calculated quantities of masterbatches along with other compounding ingredients were compounded in a lab scale two-roll mixing mill. The two-roll mill used is of roll length 13 in. and roll diameter 6 in. (Bharaj Machineries make) was operated at a friction ratio of 1:1.25. The compounding was carried out as per the formulations given in Table 2.1. The optimum cure time was evaluated on TechPro Rheotech oscillating disc rheometer (ODR) maintaining 170°C on both discs of ODR according to ASTM D-2084 standards. The compounds were cured up to their respective optimum cure time at 170°C in an electrically heated laboratory hydraulic press at 20 MPa to make approximately 2mm thick rubber sheets. The blends of EPDM and CIIR prepared were designated as ExCIy, where x is the percentage of EPDM and y is the percentage of CIIR in the blend. Thus, E80CI20 represent a blend having 80% EPDM and 20% CIIR.

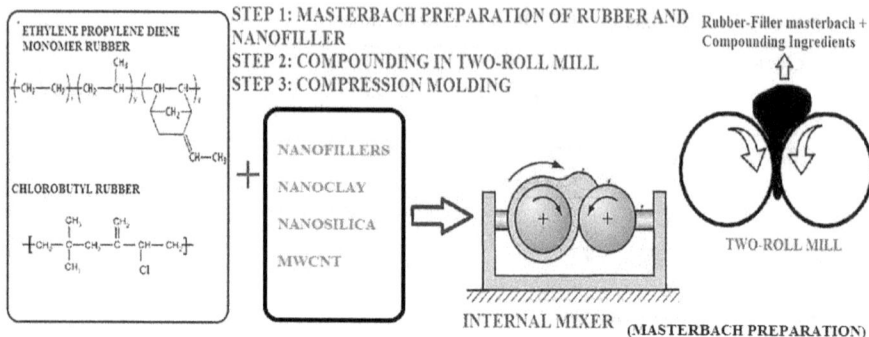

Figure 2.1 Preparation methodology for nanocomposites

The nanocomposites prepared for blends with 80% EPDM and 20% CIIR were designated as follows

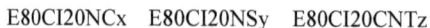

$$E80CI20NCx \quad E80CI20NSy \quad E80CI20CNTz$$

where x,y and z represents the contents of nanoclay, nanosilica and MWCNT respectively. Thus, E80CI20NC5 represents nanoclay content of 5 phr in the blend with 80% EPDM and 20% CIIR. In case of hybrid nanocomposites, the samples are designated as EP80CI20NC5NS10 representing 5 phr NC and 10 phr NS in blend with 80% EPDM and 20% CIIR.

2.4 Characterization methods

2.4.1 Cure characteristics

The cure characteristics of the different samples prepared were evaluated using TechPro Rheotech ODR (ASTM D-2084). The optimum cure time (t_{90}), cure rate, scorch time (ts_2), minimum and maximum torques (τ_{min} and τ_{max}) were determined at 170°C. The cure temperature was set at 170°C for both top and bottom discs of ODR.

2.4.2 Fourier Transform Infrared Spectroscopy (FTIR)

Attenuated total reflected FTIR spectra were recorded on Thermo Scientific Nicolet iS10 spectrophotometer at room temperature. ATR sampling attachment used is Thermo Scientific smart iTR and equipped with diamond turned reflector and infrared prism at 40° angle of mid frequency range with a resolution of 0.25 cm^{-1} in room temperature. The FTIR spectra were recorded for wavenumbers between 400 cm^{-1}- 4000 cm^{-1} and the chemical functional groups as well as their interactions were analyzed.

2.4.3 Morphological characterizations

Wide angle X-Ray Diffraction (XRD) was performed at 3°/minute in Bruker D8 ADVANCE XRD with Cu X-ray beam of wavelength 1.5406 A° to study the dispersion of layered silicate structure based on d-spacing. The interlayer distance of organoclay in the nanocomposite was determined from the position of diffraction peak in XRD diffractograms using Bragg's law.

$$n\lambda = 2d \sin\theta \hspace{5cm} (2.1.)$$

where n is an integer, λ is the wavelength, d is the d spacing and θ is the incident angle.

Transmission Electron Microscopy (TEM) analysis carried on ultra thin sections of samples prepared by cryo-ultramicrotomy was examined with JEOL-2010 transmission electron microscope (manufactured by JEOL Ltd., Tokyo, Japan). TEM analysis was carried out in bright mode at operating voltage of 200kV. The magnification of TEM micrographs were up to 15000X.

Scanning Electron Microscope (SEM) analysis carried on gold sputtered fractured surface was examined with JEOL JSM-6490 LASEM machine at an acceleration voltage of 15kV.

2.4.4 Static and dynamic mechanical properties measurement

Static and dynamic mechanical properties were evaluated with test specimens punched out of the compression molded sheets. Static mechanical properties were evaluated using universal testing machine (UTM), (Instron 3365, US - 300kN, 5kN load cell range) at a crosshead speed of 500mm min^{-1} conforming to ASTM D412 standards at room temperature. The mean value obtained for five test specimens for each composition is reported in this study. Dumbbell specimens of D type geometry (model no. DIE.D0412.0004 - 10.16 cm (4") total length, 3.32 cm (1.31") gage length, 1.57 cm (0.62") dumbbell grip width and 0.32cm (0.125") gage width) were used.

The dynamic mechanical properties were investigated with dynamic mechanical analyzer (Metravib 50N, France v 6.83). DMA was conducted in tension-compression mode in temperature sweep from -70°C to +70°C at 10Hz frequency and 0.001 dynamic strain. The test specimen had dimensions of 10mm*5.94mm*3mm (length x width x thickness). The heating rate was maintained at 5°C per minute during temperature sweep.

2.4.5 Rheological characterizations

Mooney viscosity studies were carried out on Mooney Disc Rheometre (MDR- MV 2000) at 170°C as per ASTM D1646 standards. The data extracted from MDR includes mooney viscosity (Final Value (FV) and Delta Mooney).

Rubber Process Analyzer (RPA-2000) was employed for studying Payne effect (strain sweep) and stress relaxation of nanocomposites based on EPDM-CIIR blends. Both Payne effect and stress relaxation test were carried out at 60°C as per ASTM D5289 standards. The dynamic strain conditions were maintained at 0.28 – 300% strain sweep.

2.4.6 Thermo Gravimetric Analysis (TGA)

Thermogravimetric analysis was performed in temperature range of 30°C to 700°C with nitrogen flow of 100 mLmin^{-1} at 20°C/min heating rate (SDT Q600 V20.9 machine).

2.4.7 Differential Scanning Calorimetry (DSC) analysis

DSC analysis was carried out at 10°Cmin-1 heating rate in the temperature range -80°C to 420°C in nitrogen atmosphere on DSC Q20 V24.10 Build 122 (TA instruments).

2.4.8 Solvent sorption experiments

To evaluate sorption behavior, uniform sized samples of 2cm in diameter were taken. The tested specimens were of round shape to avoid stress concentration at the edges and to obtain uniform absorption. The thickness and weight before immersion in solvent were measured using digital vernier and electronic balance respectively. Cyclohexane was chosen as the solvent for the studies based on its solubility parameter (δ) which is similar to that of EPDM and CIIR rubbers (δ_{EPDM}=16.7MPa$^{1/2}$, δ_{CIIR}=16.6MPa$^{1/2}$ and $\delta_{CYCLOHEXANE}$=16.6MPa$^{1/2}$). The diffusion experiments were performed at room temperature. The cured samples were completely immersed in glass diffusion bottles containing cyclohexane. After immersing the samples in cyclohexane, the samples were removed from the solvent at specific time intervals and weight gain was measured after removing excess solvent at the surface using filter paper until equilibrium was established. The average values obtained for five samples were reported to ensure accuracy. The mole percent uptake for the solvent, Q_t at time t was determined using the formulae

$$Q_t = \frac{M_t - M_0}{M_0 * MW} * 100 \qquad (2.2.)$$

where M_t is the mass of sample after time t of immersion, M_0 is the initial mass of the sample and MW represents the molecular weight of the solvent. The sorption isotherms were plotted with mole percentage uptake of solvent (Q_t) versus square root of time.

2.4.9 γ-Radiation test

The samples were exposed to three cumulative doses (0.5, 1 and 2 MGy) of gamma rays in ambient conditions of temperature and pressure. The gamma radiation was produced in a γ chamber from ^{60}Co source at a dosage rate of 2.97KGy/hr. The source of radiation is supplied by Board of Radiation and Isotope Technology (BRIT), Mumbai, India. The duration of radiation exposure is calculated by the formula

$$\text{Duration of radiation exposure} = \frac{\text{Cumilative doses of radiation (MGy)}}{\text{Dosage rate (kGy/hour)}} \qquad (2.3.)$$

2.4.10 Electron Spin Resonance (ESR) spectroscopy

ESR analysis was carried out to evaluate the presence of free electrons in the samples (before and after γ-irradiation). ESR spectroscopy was carried out at room temperature, resolution of 2.35μT, sensitivity of $7*10^9/0.1$ mT and at frequency range of 8.75-9.65 GHz in JEOL-JES FA200 spectrometer. ESR spectroscopy was probed at X band frequency (~9.5 GHz) to conveniently evaluate the rotational movement or dynamics of elastomeric chains and presence of free radicals embedded in rubber matrix.

Chapter 3

Studies of EPDM-CIIR blends in gamma-radiation environments[1]

3.1. Introduction

This chapter details the static mechanical properties, solvent transport characteristics and gamma radiation ageing behavior of blends of EPDM and CIIR rubbers. The blending would impart combination of properties of both rubbers and also enables to overcome the drawback of one rubber by the advantage of its counterpart [57]. EPDM rubber is a material of choice in nuclear applications due to its radiations resistance characteristics [14]. In addition, EPDM rubbers have high resistance to heat, ozone, cold temperature and moisture [100]. In nuclear fuel reprocessing, in addition to radiation, the elastomeric components have to withstand paraffinic hydrocarbons as well. The ability of EPDM rubbers to endure hydrocarbon solvents is low. This drawback can be overcome without compromising the radiation resistance by blending EPDM with a suitable polymer. Generally, blending of two polymers enhances he mechanical properties, ageing resistance and processing characteristics [54], [72], [110]. CIIR have low permeability to gas and moisture, high thermal stability and good resistance to weathering and hydrocarbon solvents. To enhance the durability of EPDM in such environments, EPDM-CIIR blends of varying compositions were developed and characterized for mechanical, solvent sorption and gamma radiation ageing behavior. This chapter also represents the spectroscopic, thermal and morphological studies of the blends prepared for evaluating the compatibility. To obtain good properties in blends, it is necessary that the blends are compatible and have minimum interfacial tension between the two polymer phases. EPDM and chlorobutyl rubber are compatible with each other as evident from their solubility parameters of $16.7 MPa^{1/2}$ and $16.6 MPa^{1/2}$ respectively %[100]. The influence of blend composition on cure kinetics, mechanical properties and solvent sorption characteristics of EPDM – CIIR blends were also detailed in this chapter.

---

[1] This chapter has been published in *Journal of Applied Polymer Science* (Neelesh Ashok *et al.*, "EPDM–Chlorobutyl rubber blends in gamma-radiation and hydrocarbon environment: Mechanical, transport and ageing behavior", *2017*, 134(33): 45195)

The optimal composition of blends with superior mechanical properties and solvent resistance were found due to synergistic effect.

In the very few studies reported on properties of EPDM – CIIR blends, the focus has been on the effect of processing parameters and compatibilization [77], [80] on mechanical [78] [76], thermal and transport properties. The effect of irradiation on the properties of these blends has not yet been published in scientific literatures. The optimized blends were irradiated with gamma rays at cumulative doses upto 2MGy. Based on spectroscopic, mechanical properties, thermal analysis and sorption coefficients, and the blend containing 80% EPDM was found to have superior retention of properties after γ-irradiation.

3.2 Results and discussions

3.2.1 Cure characteristics

The rheograms of EPDM-CIIR Blends obtained at 170°C are shown in Figure 3.1(a) and the cure characteristics of EPDM – CIIR blends are tabulated in Table 3.1. The notation ts_2 and tc_x represent scorch time and the time taken to reach x% of maximum torque. As evident from Table 3.1, the optimum cure time (t_{90}) and scorch time (ts_2) increased with increasing CIIR content in the blend.

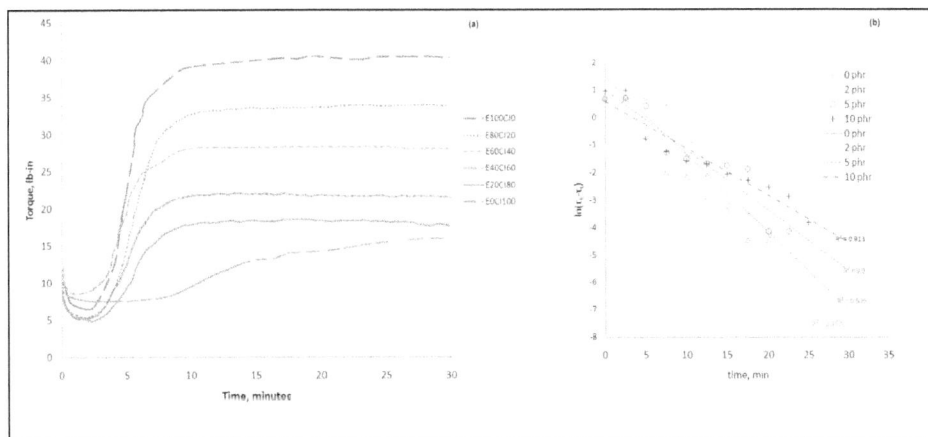

Figure 3.1. Cure characteristics of EPDM-CIIR blends (a) rheogram and (b) cure kinetics

Sui [170] have reported that in sulfur curative systems there is preferential migration of curative occurring to the EPDM phase as compared to the CIIR phase. This result in larger number of cross- link interfaces in EPDM and consequently the time taken for optimum cure is reduced in blends containing higher percentage of EPDM. The increasing value of the difference between maximum and minimum torque of the blends with increasing EPDM content is also evidence to the increase in extent of cross- linking. The torque vs. time data plots (rheograms in Figure 3.1(a) with representative values given in Table 3.1) were analyzed to determine the order of cure reaction.

Table 3.1. Cure characteristics of EPDM-CIIR blends

Sample	Min Torque τ_{min}, Nm	Max. Torque τ_{max}, Nm	$\tau_{max} - \tau_{min}$, Nm	ts_2 minutes	t_{20} minutes	t_{50} minutes	t_{90} minutes	CRI minute^{-1}
E0CI100	0.40	1.38	0.98	7.20	6.15	11.5	31.0	4.20
E20CI80	0.56	1.71	1.29	1.36	1.40	2.38	22.4	4.76
E40CI60	0.56	2.08	1.56	1.36	1.36	2.27	12.1	9.30
E60CI40	0.52	2.39	1.83	1.33	1.33	2.23	9.12	12.8
E80CI20	0.42	2.47	1.91	1.30	1.29	2.22	7.55	16.0
E100CI0	0.56	4.02	3.46	1.08	1.21	1.47	5.39	23.2

The cure rate index (CRI), given by equation (3.1.) is a measure of the fast curing nature of rubber compounds

$$CRI = \frac{100}{t_{90} - t_{s2}} \qquad (3.1.)$$

The CRI of the blend increased as a linear function of EPDM content in the blend ($R^2 = 0.953$). Cure kinetics relating degree of cross linking (α) and vulcanization time (t) of rubber compounds is given by equation (3.2)

53

$$\frac{d\alpha}{dt} = k\,(1-\alpha)^n \qquad\qquad (3.2.)$$

where $d\alpha/dt$ is the vulcanization rate, k is specific rate constant at temperature T and n is the order of the cure reaction . For a first order cure reaction, equation (2) modifies to the form given in equation (3.3.) relating degree of cross linking α to the parameters of the oscillating disc rheometer study

$$\alpha = \frac{\tau_t - \tau_0}{\tau_h - \tau_0} \qquad\qquad (3.3.)$$

where τ_0, τ_t and τh are the torque values at time zero, at time t and at the end of curing, respectively. For first order kinetics (n = 1), equations (3.2) and (3.3) can be combined and integrated with respect to t to give

$$\ln\,(\tau_h - \tau_t) = k(T)\,t + \ln\,(\tau_h - \tau_0) \qquad\qquad (3.4.)$$

To verify the compliance of experimental data to first order kinetics, $\ln\,(\tau_h - \tau_t)$ was plotted against curing time t (Figure 3.1 (b)) and fitted to the linear model given in equation (3.4.). The regression coefficients (R^2) for all the blends were found to be greater than 0.9 and first order kinetic model was found to be appropriate to describe the cure reaction of EPDM – CIIR rubber blends.

3.2.2 Fourier Transform Infrared (FTIR) analysis

FTIR spectra is an effective tool in detecting the functional group present in the sample and was used in this work to evaluate possible chemical interactions between the rubbers in the blends [171][172]. The identification of the type bonds present in the blends gives a picture of the components present in the blends. When the molecules absorb infrared radiation, the absorbed energy causes an increase in amplitude of vibrations of the bonded atoms which are already vibrating about their equilibrium position. A particular type of bond between a certain pair of atoms will have a characteristic frequency of vibration which is expressed as wave number in infrared spectroscopy.

Figure 3.2(a) shows the FTIR spectrum of 100% EPDM rubber compound. Peaks observed in the spectrum correspond to C-H stretching (2919cm^{-1} and 2850cm^{-1}), C=C stretching (1500cm^{-1} and 1558cm^{-1}) and CH_2 and CH_3 angular deformation (1370 cm^{-1} and 1460 cm^{-1}) respectively. The peak found at 912cm^{-1} corresponds to =CH-CH_2- which characteristic to EPDM rubber [173], [174]. Figure 3.2(b) shows the FTIR spectra of 100% CIIR rubber compound where the sharp and narrow band corresponding to wave number 2950 cm^{-1} represents CH bond. The peak corresponding to C-Cl bond is observed in wave number of 690 cm^{-1}.

In the case of EPDM CIIR blend, hydrogen bonding in wave number range of 2500 cm^{-1} to 4000 cm^{-1} was seen in FTIR spectra of E60CI40 and E80CI20 seen in Figure 3.2(c). It is been reported in several literature that the increase in hydrogen bond is an indication of blend compatibility. [55],[173]. Hence, it may be inferred that these blends are compatible.

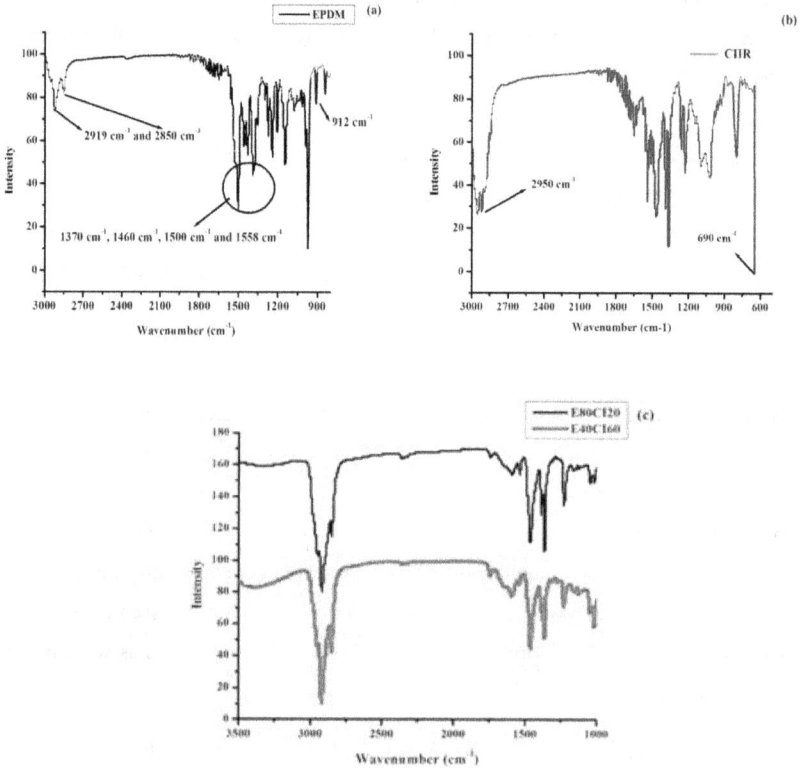

Figure 3.2. FTIR spectra of (a) EPDM rubber (b) CIIR rubber and (c) EPDM-CIIR blends

3.2.3. Morphology

SEM analysis was used to investigate the microstructure of tensile fracture surfaces of the blends of various compositions [170]. The SEM micrographs of the fractured surfaces of EPDM - CIIR blend of different compositions are depicted in Figures 3.3(a)–(c). The fractured surface of E40CI60 blend composition as shown in Figure 3.3(a) has internal rupture on the surface which depicts lack of mechanical strength during tensile test. A distinct phase separation and non-homogeneity was clearly visible in this case indicating poor compatibility. The phase separated morphology of E40CI60 indicates immiscibility, which was indicated by the reduction in mechanical properties. This observation supports the observation found in infrared spectroscopic analysis. The fractures surface on blends E60CI40 and E80CI20 as shown in Figure 3.3(b) and 3.3(c) was smooth and homogeneous confirming compatibility and supporting FTIR measurements. In the blend containing 80 % EPDM and 20% CIIR (Figure 3.3(c)), homogeneity improved mechanical properties. Some of ingredients used for compounding were also observed.

Figure 3.3 SEM micrographs of fractured surfaces of (a) E40CI60 (b) E60CI40 and (c) E80CI20

3.2.4 Differential Scanning Calorimetry (DSC)

Differential scanning calorimetry was employed to evaluate the glass transition temperature (T_g) of the EPDM – CIIR blends. A representative DSC thermogram of E80CI20 is depicted in Figure 3.4. Analyses of thermograms show that the glass transitions are -62.86°C for 100% EPDM - 37.2°C for 100% CIIR and -42.86°C for E80CI20. For the blend, the glass transition temperature lies between the two polymers. Along with this observation, the absence of two significant peaks confirms that the blend formed is miscible as confirmed by the spectroscopic analysis and morphology studies.

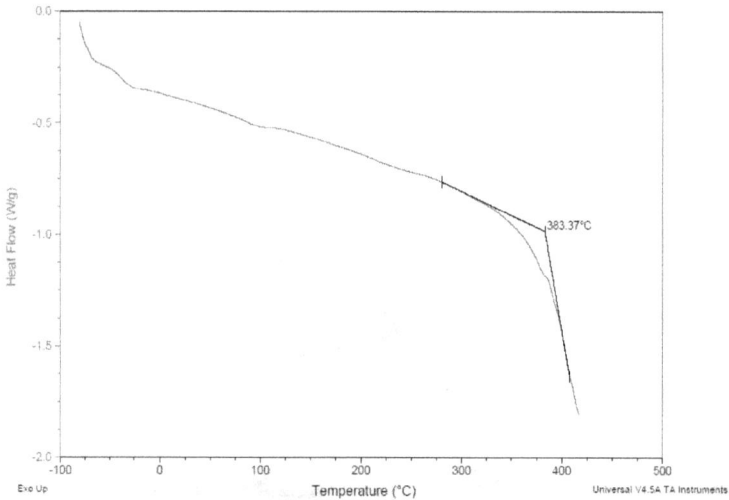

Figure 3.4 DSC thermo gram of E80CI20

3.2.5 Mechanical properties

The mechanical properties such as tensile strength and hardness of the blends were evaluated as a function of blend composition (CIIR content) and summarized in Figure 3.5 (a-c). The tensile strength of blends containing 20% and 40% CIIR is higher than that of neat EPDM and CIIR blends. This is due to the synergistic behavior of blends that resulted in better packing of polymer chains [175]. The compatibility and homogeneity of the blends at these compositions resulted in improved tensile strength. Spectroscopic, thermal and morphological analysis discussed in the previous section confirmed this observation. The tensile strength has been found low for the blend constituents having EPDM content of 40 percent and less due to immiscibility at the interface and non – homogeneity.

The Shore A hardness was measured using shore durometer by indenting a rigid ball into the rubber specimen as per ASTM D2240 standards at ambient conditions and the average of 10 measurements was reported in this study. The blend containing 20% CIIR has the highest hardness, confirming the synergistic contribution of both EPDM and CIIR to blend properties. These results conform to those obtained in morphological and thermal analysis. At higher CIIR contents, the hardness was found to decrease with increasing CIIR content. In sulfur curative systems, the preferential migration of curative to the EPDM phase as compared to the CIIR phase and the resulting larger number of cross- link interfaces in EPDM phase was responsible for higher hardness in blends having higher EPDM content.

Tensile strength after heat ageing was also higher for blends containing 60 and 80 percentage of EPDM. This may be attributed to the heat ageing resistance property of EPDM rubbers. It has been found that pre-curing of CIIR to an optimum low level and followed by blending with EPDM paves an effective method of obtaining optimum cross link density in both the elastomeric phase and at the interfaces thereby enhancing mechanical properties significantly.

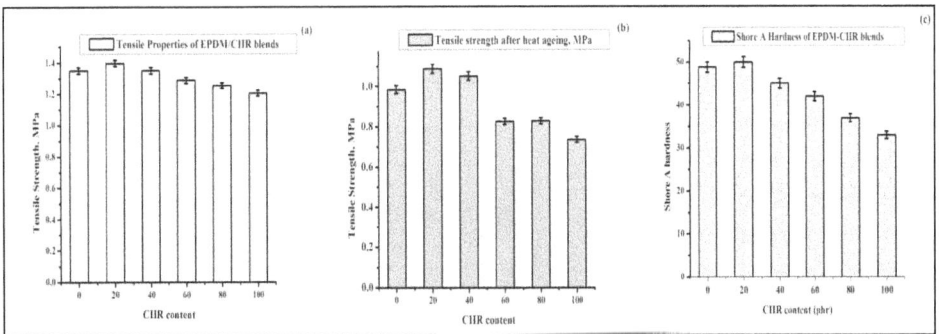

Figure 3.5 (a) Mechanical tensile strength (b) Tensile strength after heat ageing and (c) Shore A hardness of EPDM-CIIR blends

3.2.6. Solvent Sorption Behavior and Transport Properties

Transport properties of the polymeric blend can be explained by sorption, diffusion and permeation phenomena. The diffusion through a polymeric material occurs due to random vibration of individual molecule. The permeation of solvent depends on the concentration of available space in the polymer that is large enough to accommodate the penetrant molecule, size of the penetrant, temperature, polymer segment mobility, reaction between solvent and the matrix etc. Therefore the studies on transport process can be emphasized to understand the interfacial interaction and morphology of the system.

The diffusion coefficient of a polymeric sample immersed in an infinite amount of solvent can be calculated using the equation [89], [94], [95], [176]

$$\frac{Q_t}{Q_\infty} = 1 - \left(\frac{8}{\pi^2}\right) \sum_{n=0}^{n=\infty} \frac{1}{(2n+1)^2} \exp\left[-D(2n+1)^2\pi^2 \frac{t}{h^2}\right] \tag{3.5.}$$

where Q_t is the mole percent uptake for solvent at time t, Q_∞ is the mole percent uptake for solvent at equilibrium swelling, t is the time, h the initial thickness of the sample, D the diffusion coefficient and n is an integer. From equation (3.5), it can be seen that a plot of Q_t versus $\sqrt{t}$ is linear at short time interval and D can be calculated from the initial slope. For short time limit, the equation 3.5 can be modified as

$$\frac{Q_t}{Q_\infty} = \frac{4}{h} \left(\frac{D}{\pi}\right)^{1/2} t^{1/2} \tag{3.6.}$$

The mole percent uptake of the solvent, Q_t, at time t was determined using the formula

$$Q_t = \frac{M_t/M_0}{MW} \times 100 \tag{3.7.}$$

where M_t is the mass of solvent absorbed after time t of immersion, M_0 the initial mass of the sample and MW is the molecular weight of the solvent. The sorption isotherms (mole percent uptake of solvent as a function of time ½) for various blend compositions at 30°C calculated from experimental values using equation (3.7) are plotted in Figure 3.6. The Q_t vs $\sqrt{t}$ curve showed two distinct regions - an initial steep region with high sorption rate due to large concentration gradient and a second region exhibiting reduced sorption rate that ultimately reaches equilibrium sorption. The sorption and equilibrium for EPDM was found to be highest and that of pure CIIR is lowest. The higher polarity of CIIR is responsible its lower solvent uptake. Consequently, the solvent uptake of the blends progressively decreased with increasing CIIR content.

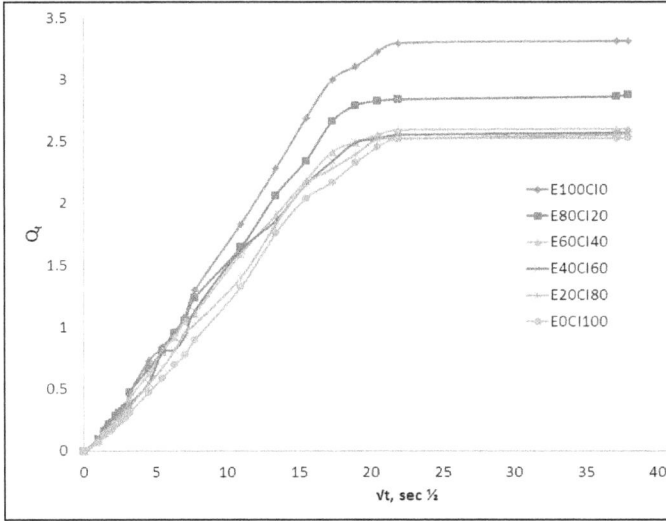

Figure 3.6 Sorption isotherms for EPDM-CIIR blends at 30°C

In cross linked polymers, the polymer segments between the cross-links take up the solvent, leading to swelling of the polymer. The initial solvent uptake for all compositions of the blend is high due to high concentration gradient for diffusion of the solvent. Subsequently, as the gradient decreased, the solvent uptake decreased and equilibrium was attained. During solvent sorption, the free energy of mixing is responsible for the penetration and consequent dilution of the polymer, resulting in increase in entropy. As the solvent uptake increases, the polymer segments elongate under swelling resulting in an elastic force that opposes the deformation caused by swelling. At equilibrium, the two forces balance and a steady state was attained [79] [177].

The swelling behavior of composites can be quantified from sorption data in terms of swelling coefficient (β), cross link density (υ) and molecular weight of the polymer between the cross-links calculated using the equations (3.8) to (3.10).

$$\beta = \frac{(M_\infty - M_0)}{M_0} \times \rho_s^{-1} \qquad (3.8.)$$

60

$$\upsilon = 1/M_c \tag{3.9.}$$

$$M_C = -\frac{\rho_P V_S \varphi^{1/3}}{\ln(1-\varphi)+\varphi+\chi\varphi^2} \tag{3.10.}$$

where M_o , M_∞ , ρ_s , ρ_P, V_s , χ and φ are mass of the sample before swelling, mass of sample at equilibrium swelling, density of the solvent, density of the polymer, molar volume of the solvent, interaction parameter and volume fraction of rubber in the solvent-swollen filled sample respectively [177]

The diffusion phenomenon across a section of the polymer can be described by Fick's first law of diffusion. Molar flux (J) defined as the moles of solute passing per unit time through a plane of unit area normal to the direction (x) diffusion is proportional to the concentration gradient (dc/dx).

$$J = -D \, dc/dx \tag{3.11.}$$

where D is the diffusivity or diffusion coefficient. For polymeric systems, sorption data can be used to calculate diffusivity and equation (3.6) can be rewritten as

$$D = \pi \left[\frac{h\theta}{4Q_\infty} \right]^2 \tag{3.12.}$$

where h is the thickness of the sample, θ is the slope of the initial linear portion of the sorption isotherm (up to 50% equilibrium solvent uptake) and Q_∞ is the molar percentage of the solvent uptake at equilibrium. Sorption coefficient (S) is the ratio of solvent uptake at equilibrium to initial mass of sample. Permeability or permeation coefficient (P) is the product of diffusion and sorption coefficients (P = D x S).

Table 3.2 summarizes the transport coefficients for the various blends. The coefficients of diffusion, sorption and permeation decreased as a function of increasing CIIR content. The polar nature of CIIR reduced the mobility of the non-polar solvent in the polymer blend. It may be concluded that blending EPDM with CIIR increased the resistance to hydrocarbon permeation and EPDM – CIIR blends will be suitable for applications where the rubber is exposed to hydrocarbon environment. The average values of five samples were reported in Table 3.2 as discussed in previous chapter.

61

Table 3.2 Swelling coefficient, crosslink densities and transport coefficients of EPDM/CIIR blends

Sample	Swelling coefficient (β) (cm^3/g)	Molar mass between crosslinks (M_c) (g/mol)	Crosslink density x 10^4 (gmol/cm$^{3)}$	Diffusion coefficient (Dx10^7) (m^2/s)	Sorption coefficient (S) (g/g)	Permeability coefficient (P x10^7) (m^2/s)
E100CI0	3.58	2568	3.89	3.97	2.79	11.07
E80CI20	2.04	2468	4.05	3.51	1.59	5.58
E60CI40	1.94	1990	5.02	2.83	1.49	4.22
E40CI60	1.78	1965	5.09	2.72	1.38	3.75
E20CI80	1.77	1908	5.24	2.65	1.36	3.60
E0CI100	1.73	1876	5.33	2.56	1.35	3.45

3.2.7 Gamma-irradiation studies

Based on mechanical properties and considering miscibility of the blends based on FTIR analysis and morphological studies, it can be concluded that optimum properties in blends were confined to blend compositions of 60 and 80 percentages of EPDM content. Hence further studies on effect of gamma irradiation were carried out with blends E80CI20 and E60CI40 and compared with EPDM (E100CI0).

3.2.7.1 FTIR Analysis of Irradiated Samples

The effect of irradiation was studied by comparing FTIR spectra of blends subjected to irradiation with that of non-irradiated ones [40]. Conformation on chemical changes caused by irradiation can be investigated by analyzing formation of new bonds or functional groups using FTIR. Figure 3.7 shows the comparison of FTIR spectrum of irradiated E80CI20 with that of non-irradiated E80CI20.

In rubbers, the chain scission of polymer and free radical generation [178] is induced as an effect of irradiation. The generated free radicals interact with molecular oxygen to form peroxyl radicals and irradiation oxygen. Compounds containing carbonyl and hydroxyl functional groups like carboxylic acids, ketones, hydroperoxides corresponds to oxygenated products. In Figure 3.7, the IR absorbance peak at 1622 cm^{-1} in the irradiated blend of E80CI20 confirms the formation of carbon-carbon double bond (C=C stretching 1620cm^{-1}to 1680cm^{-1}) due to the effect of cross linking [179] with simultaneous decrease in C-Cl bonds. This confirms the chemical changes that occurred due to irradiation [180]. The chemical interactions seen in FTIR spectrum is due to radiation induced crosslinking and/or chain scission [181].

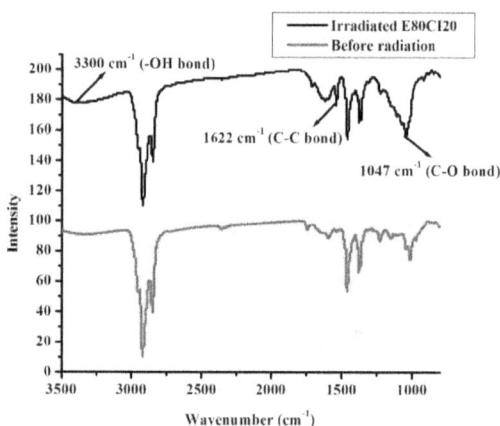

Figure 3.7 Comparison of FTIR spectra of irradiated and non-irradiated E80CI20

3.2.7.2 Thermo Gravimetric Analysis (TGA)

TGA observes the weight loss and rate of variation of mass with respect to time or temperature. It is primarily used to determine the thermal and or oxidative stabilities of the material. This technique has the ability to analyze the material that exhibit either weight gain or loss due to decomposition, oxidation or loss of volatiles like moisture content. Thermal degradation occurs when exposed to gamma irradiation as a result of formation of chain scission, cross link scission, cross links [181]. The extent of cross linking and degradation undergone by a polymer depends on its structural characteristics and presence of initiators or sensitizers. The thermal degradation behavior of gamma irradiated E80CI20 were determined from TGA data. The thermal properties of E80CI20 (initial, peak and final degradation temperatures) irradiated at all doses are tabulated in Table 3.3. The irradiation has significant on the thermal degradation onset temperature which is due to radiation induced effects. It can also be inferred that though irradiation have lesser impact on temperature at which maximum degradation takes place, irradiated blends had greater percentage of weight loss and rate of change of mass with respect to time. As a representation, TGA analyses of E80CI20 were considered in this study, similar trends was observed for E60CI40 and E100CI0. This observation is in concurrency with static mechanical behavior and solvent sorption characteristics of irradiated blends.

Table 3.3 Thermal properties of irradiated E80CI20

	E80CI20			
Thermal properties	Un-irradiated	0.5MGy	1MGy	2MGy
Onset of degradation temperature (°C), T_i	202	210	225	223
Peak degradation temperature (°C), T_d	453	453	456	454
Terminal degradation temperature (°C), T_o	485	486	489	481
% char	92.6	92.9	95.4	93.9

3.2.7.3 Mechanical properties after γ-irradiation

Most obvious effect brought about by irradiation on the blends due to their cross linking and / or degradation is observed on the tensile strength of the material. The influence of γ-radiation affects the polymer in two ways – chain scission which results in reduced tensile strength and increased elongation at break or cross linking which increases tensile strength but reduces the elongation at break. The resistance towards radiation in aliphatic polymers is dependent upon levels of unsaturation and substitution, whereas aromatic polymers are generally resistant to radiation effects. To obtain the influence of radiation on blends, comparative study of tensile properties of blends exposed to varying cumulative doses of gamma radiation is presented in this work. In E100CI0 and E80CI20, the tensile strength increased as a function of radiation doses of 0.5 and 1.0 MGy which is illustrated in the Table 3.4. The mechanical strength declined for all blends when subjected to high dose of gamma rays (2 MGy) [182]. As observed from FTIR analysis, the rubber undergoes cross linking upon irradiation and the increase in cross-linking is responsible for increase in tensile strength [183]. The elongation at break was found to decrease with the radiation dose for EPDM and all blends due to more cross linkage formations on the structure that deteriorates internal chain mobility and elongation [183].

Upon irradiation of the rubber specimens, it was observed that there was an increase in modulus. Modulus is a material property and it may be inferred that the brittleness in the blend subjected to irradiation was the result of molecular modifications, i.e., cross-linking as seen in FTIR spectroscopy. In case of E60CI40, as the CIIR content increased, degradation also increased as CIIR have less radiation resistances compared with EPDM, thereby lowering mechanical strength and crosslink density.

The irradiation effect on mechanical properties was illustrated with the aid of percent retention rate given in equation 3.12. Percent retention rate can be defined as percent ratio of tensile property value of irradiated blend to that of non irradiated blend [21]

$$\% \text{ retention TS} = \frac{\text{Tensile strength of irradiated blend}}{\text{Tensile strength of non-irradiated blend}} \times 100 \qquad (3.12.)$$

The increase in percent retention rate due to relative increase in tensile strength attained by irradiated samples with respect to that of un-irradiated ones, attributed to the occurrence of cross linking taken place during irradiation [59]. The dominance of radiation induced chain scission or degradation at higher radiation dose led to decline in percentage retention rate. The specimens were found to lose elasticity in blend compositions when exposed to higher irradiation dose of 2.0MGy. At higher irradiation doses, degradation reaction occurs leading to cracks initiation that consequently lowers the mechanical strength. The lower tensile strength in samples exposed to larger radiation doses can be attributed to radiation induced chain scission or oxygenative degradation. The formation of oxygenated bonds after irradiation as evident from FTIR spectra can be corroborated with this behavior.

As seen in FTIR analysis of irradiated EPDM-CIIR blends, the effect of γ irradiation is to break the carbon-halogen bond which is weaker compared to carbon-carbon and carbon-hydrogen bond to generate organic free radical [184] which changes the mechanical properties. These free radicals generate intermediates for subsequent processes like oxidation and further propagate to form chain reactions. Under the effect of gamma radiation, due to chain scission at higher doses, significant degradation takes place as CIIR is blended with EPDM. The presence of halogen in CIIR induces lower material resistance to radiation action and scission takes place faster. It can be concluded that increasing radiation dose results in significant loss in properties of sulfur vulcanized compounds. The tensile strength of a specimen is indicative of the failure property [21], [47], [178] and from Table 3.4, it was observed that the percent retention in tensile strength of irradiated samples increased with the increase in CIIR content. From the mechanical properties tabulated in Table 3.4, it was delineated that blend with 80% EPDM content was found to have least changes in properties (percent retention) after irradiation. Compared with E60CI40, the least deviation in the percentage retention in tensile strength can be attributed to presence of more EPDM content, which provides radiation resistance.

Table 3.4 Tensile strength of irradiated blends

Blend	Before radiation	Low (0.5 MGy)	Medium (1 MGy)	High (2 MGy)	Percentage retention in TS		
					0.5 MGy	1 MGy	2 MGy
E60CI40	1.34±0.06	1.39±0.05	2.07±0.09	2.06±0.10	103	156	153
E80CI20	1.39±0.07	1.38±0.05	2.08±0.11	2.06±0.11	99	149	148
E100CI20	1.37±0.04	1.48±0.06	2.01±0.10	2.13±0.08	108	147	155

3.2.7.4 Transport coefficients of irradiated blends

In order to evaluate swelling coefficient and cross link density, the samples of irradiated rubber blends of varying CIIR content (0, 20 and 40 percent) was immersed in cyclohexane at room temperature. Solvent uptake measurement is a direct approach to evaluate cross link density. The polymeric chain mobility gets diminished on cross linking thus reduces the transport/sorption. Molecular transport of the cyclohexane has been studied for the irradiated samples and results are tabulated in Table 3.5. In irradiated specimens, it can be seen that the swelling index decreased as cross linking increased for blends. For 40% CIIR it was found that percentage change in swelling coefficient increased owing to degradation caused by irradiation [185].

Table 3.5 Swelling coefficient and crosslink density of irradiated blends

Samples	Swelling coefficient (β) (cm^3/g)				Percentage change with unirradiated		
	Before radiation	Low (0.5 MGy)	Medium (1MGy)	High (2MGy)	Low (0.5 MGy)	Medium (1MGy)	High (2MGy)
E60CI40	1.94	1.25	1.22	1.24	35.6	37.1	36.1
E80CI20	2.04	1.35	1.31	1.33	33.8	35.8	34.9
E100CI0	3.58	1.41	1.38	1.35	60.6	61.4	62.3
Samples	Crosslink density x 10^4 ($gmol/cm^3$) (v)				Percentage change with unirradiated		
	Before radiation	Low (0.5 MGy)	Medium (1MGy)	High (2MGy)	Low (0.5 MGy)	Medium (1MGy)	High (2MGy)
E60CI40	5.02	11.3	15.0	14.8	125	199	195
E80CI20	4.05	5.03	9.89	10.1	24.2	144	149
E100CI0	3.89	5.07	9.76	9.89	30.3	151	154

67

3.3 Conclusions

This chapter contains detailed investigations on the mechanical, thermal, and sorption properties of EPDM- Chlorobutyl rubber (CIIR) blends with focus on application in gamma irradiation and hydrocarbon environment. These blends were prepared in various compositions based on different blend ratios and evaluated for various properties. The cure characteristics were analyzed and it was found that the scorch time and cure time of the blends increased with increasing CIIR content whereas the cure rate index decreased with increasing CIIR content. The cure kinetics of the blends was found to follow first order kinetics. The EPDM – CIIR blends were found to be miscible as ascertained by spectroscopic studies, differential scanning calorimetry and scanning electron microscopy. Mechanical properties varied with blend composition and highest tensile strength was obtained for blends containing 20% and 40% CIIR which is attributed synergistic behavior of blends. These results have been corroborated with morphological and FTIR studies.

The transport characteristics of the blends in cyclohexane were also evaluated and transport coefficients such as sorption, diffusion and permeation was found to decrease with increasing CIIR content owing to the increase in polar nature of the polymeric blend. The swelling coefficient and cross-link density calculated from the sorption data verified this trend. Effect of irradiation on the blends was studied by exposing blends to gamma radiation from ^{60}Co source at dosage rate of 2.67 KGy/hr for cumulative doses of 0.5 MGy, 1 MGy and 2 MGy. In terms of percent retention rate of mechanical properties, blend containing 80% EPDM and 20% CIIR was found to be better. The optimal blend composition of 80% EPDM and 20% CIIR content based on least change in properties after irradiation was used for further studies in next chapters. The changes in mechanical, thermo gravimetric and solvent sorption of the irradiated blends have been corroborated with FTIR analysis.

Chapter 4

Organo-modified layered silicate nanocomposites of EPDM-CIIR rubber blends for gamma radiation environments[2,3]

4.1 Introduction

The properties of elastomer blends can be further improved by suitable additives like fillers. Elastomer – nanocomposites have attracted great interest in the last two decades due to their fascinating properties. In conventially used elastomer micro composites, the properties are dictated by the bulk properties of matrix and reinforcing filler. But in case of nanocomposites, in addition to the matrix and filler properties, the nanofiller dispersion and polymer – filler interfacial interactions are also significant in determining properties. Interfacial structure of the nanocomposite differs from that of bulk structure and imparts enhancement of mechanical, thermal, electrical, optical, barrier, heat resistance, flame retardancy properties even at low filler content (less than 10wt%). Several researchers have reported improvements in mechanical, thermal, barrier and degradation resistant properties in elastomers reinforced with layered silicates (nanoclay). In this chapter, blends of EPDM rubber and CIIR were reinforced with organo-modified layered silicate (nanoclay) to enhance their performance in radiation as well as hydrocarbons environments. From the studies on the effect of blend ratios of EPDM and CIIR rubbers given in the previous chapter, it was revealed that blends with 80 % EPDM and 20% chlorobutyl rubber had higher mechanical properties, lower hydrocarbon transport coefficients and enhanced gamma radiation ageing characteristics. Incorporation of nanoclay in E80CI20 would further enhance the durability in radiation and hydrocarbon environment.

---

[2] A part of this chapter is published in *Journal of Composite Materials* (Neelesh Ashok *et al.*, "Organo-modified layered silicate nanocomposites of EPDM–Chlorobutyl rubber blends for enhanced performance in radiation and hydrocarbon environment", ***2018***, 52(23): 3219-3231)

[3] Another part of this chapter is published in *Composite Interfaces* (Neelesh Ashok *et al.*, "Nano-reinforcement mechanism of organomodified layered silicates in EPDM/CIIR blends: experimental analysis and theoretical perspectives of static mechanical and viscoelastic behavior", ***2020***, DOI: 10.1080/09276440.2020.1736879)

The morphology and physico-chemical interactions were evaluated by XRD, TEM and FTIR and correlated to the enhancement in mechanical properties. This chapter also focuses to explore the influence of nanoclay on the dynamic mechanical properties of blends. The applicability of various analytical models to predict the static and dynamic modulus is explored in this chapter. Mooney Rivlin plots, filler-rubber interactions and stress relaxation of EPDM-CIIR nanocomposites are also explained in this chapter. The nano reinforcement parameters, calculated from DMA data are discussed in detail.

4.2 Results and discussions

4.2.1 Cure Characteristics

From the torque vs time data obtained from cure studies using ODR, optimum cure time, scorch time and cure rate index (CRI) were evaluated and analyzed. The cure characteristics of EPDM-CIIR nanocomposites are summarized in Table 4.1. It was evident that scorch time (t_{s2}) of the nanocomposites were higher than that of the unfilled blend. Scorch time is a measure of the processing safety and it may be inferred that the formation of intercalated structures of nanoclay prevents premature curing of the nanocomposites. The intercalation of nanoclay between the rubber chains restricted the movement of free radicals for cross-linking. This resulted in hindering curative - matrix interaction that delayed initial curing [95]. The optimum cure time (t_{90}), is a measure of cure time. t_{90} also escalated with increase in nanoclay content owing to increase in filler- rubber interaction arising from exfoliation and intercalation of layered silicate in the rubber blend. As a consequence, molecular motion of polymer chains were restricted thus imposing additional resistance to form cross-links. At higher nanoclay content (beyond 7.5 phr), there was no further increase in optimum cure time and scorch time due to agglomeration of layered silicates [95]. Another useful parameter is CRI to estimate vulcanization characteristics. An increase in CRI with nanoclay content confirmed that organomodifier in the nanoclay supported the activation of cure reaction. Torque is a measure of the shear modulus while $\Delta\tau$, the difference between maximum (τ_{max}) and minimum torque (τ_{min}) is a measure of extend of cross-linking. τ_{max} τ_{min} and $\Delta\tau$ increased with increasing nanofiller content owing to increase in interaction between nanoclay and rubber [186].

Table 4.1. Cure characteristics of EPDM-CIIR nanocomposites at 170°C

	E80CI20	E80CI20NC2.5	E80CI20NC5	E80CI20NC7.5	E80CI20NC10
Scorch time t_{s2} (min)	1.30	1.37	1.48	1.58	1.47
Optimum cure time t_{90} (min)	7.55	7.27	7.42	8.27	8.28
CRI (min^{-1})	16.0	16.9	16.9	14.9	14.7
Minimum Torque τ_{min} (N-m)	2.47	2.67	4.70	7.20	10.0
Maximum Torque τ_{max} (N-m)	4.67	7.03	11.5	14.0	20.7
$(\tau_{max}-\tau_{min})$ (N-m)	2.22	4.36	6.80	6.80	10.7

4.2.2 Attenuated FTIR analysis

The FTIR spectra of the EPDM-CIIR nanocomposites were analyzed to understand the occurrence of chemical interactions between the dispersed layered silicates and the rubber matrix. FTIR spectroscopy is an effective tool used by researchers for investigating chemical interactions [187]–[190]. FTIR spectra for organo-modified nanoclay (layered silicate) reinforced EPDM-CIIR blends, unfilled blend and nanoclay are plotted in Figure 4.1. In FTIR spectrum of E80CI20, peak observed at wavenumber 2850 cm^{-1} represents symmetric stretching of –CH bond in –CH$_2$ group. The presence of Al-OH and Si-O-Si functional groups in nanoclay was evident from broad peaks around 3633 cm^{-1} and 1046 cm^{-1} respectively. The peak observed at 1642 cm^{-1} can be attributed to the possible presence of moisture in the polymer [191]. The functional groups present in nanoclay and neat blends were visible in the nanocomposites also.

71

New peak observed at 2920 cm^{-1} in nanoclay reinforced blends indicated presence of asymmetric stretching of –CH bonds in –CH$_2$ group and –NH bond [191]. In EPDM-CIIR nanocomposites, new asymmetric –CH stretch bond would have formed as a result of chemical interaction between organomodified layered silicates (OMLS) and rubber. A broad peak observed around 3365 cm^{-1} band in OMLS reinforced blends were formed from exchangeable protons from alcohol, amine or acidic groups present in organomodifer of OMLS. The peak found at 1454 cm^{-1} in nanocomposites can be attributed to aliphatic –CH bond in rocking bonds of methylene (-CH$_2$) group [192] present in organomodifer [193]. In E80CI20NC10, the depth of peaks at 1454 cm^{-1} (C-H bending) arising from CH$_3$ group in nanoclay, 1031 cm^{-1} (stretching and flexural vibrations of Si-O-Si) and 3500 cm^{-1} (–OH groups) increased due to the higher OMLS content.

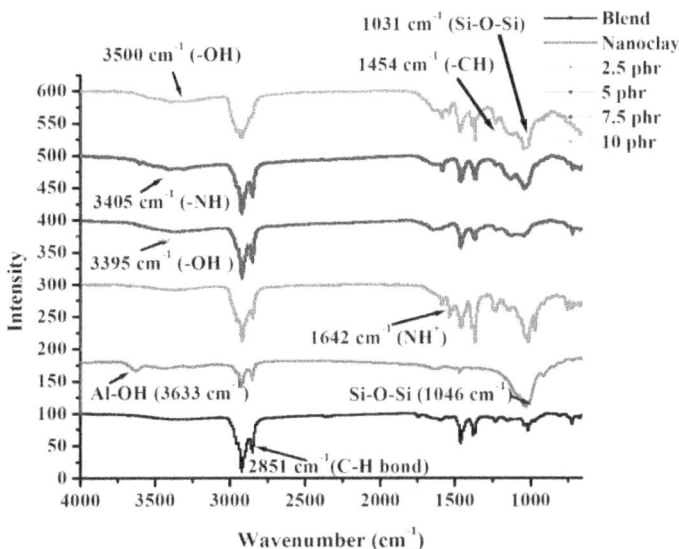

Figure 4.1 FTIR spectra of EPDM-CIIR nanocomposites

4.2.3 Morphology

The enhancement of properties in nancomposite is primarily dependent on degree of dispersion of nanosized filler in polymer matrix. Ishida *et al.* studied the effect of polymer intercalation into layered silicates and its effect on nanocomposite properties using XRD [192], [194], [195]. The variation in breadth, intensity and position of the XRD spectra gives an insight about the structural characteristics of nanoclay. XRD spectra also provide information on the state of intercalation, dispersion and exfoliation of layered silicates in the matrix. The layered structure in nanoclay gives rise to a peak in the XRD spectra. The d-spacing or the spacing between the layers of OMLS is calculated from Bragg's law $n\lambda = 2d\sin\theta$ where n, λ, d and θ represent the order, wavelength of incident x-ray, d spacing and the angle at which the peak appears in the spectrum respectively. In the OMLS (or nanoclay) used in this work, the peak of intensity appeared at $2\theta = 4.34°$ and the d spacing was calculated to be 2.015 nm. The XRD spectra of OMLS or nanoclay and nanocomposites are plotted in Figure 4.2.

Compared to nanoclay (nanomer 1.34nm), the spectra of the nanocomposites do not show any prominent peaks. This implies that the layers of the nanoclay could have delaminated. However, a closer look at spectra shows smaller peaks of lower intensity for the nanocomposites. A negligible peak was observed for E80CI20NC2.5, which indicates intercalation of rubber chains in between layered galleries. The layers of nanoclay would have delaminated and exfoliated between the rubber chains. There are no identifiable peaks for EPDM-CIIR blends up to 5 phr nanoclay.

In blends with 7.5 phr and 10 phr nanoclay content smaller peaks of intensity 1000 and 1400 was observed at $2\theta = 6.81°$ and $2\theta = 6.86°$ respectively. The smaller peaks arise from the presence of small agglomerates of OMLS that retains the layered structure. The shift in the position of peak in nanocomposites implies that there was change in d-spacing of layered silicate. The d-spacing of layered silicates in the nanocomposites was calculated to be 1.29 nm.

It has been postulated that alkyl group in organo-modified clay participates in curing reaction resulting in collapse of layers of nanoclay and decrease in d-spacing [186]. During cure reaction, the zinc-sulfur accelerator complex formed "extracts" the organo-modifier group in nanoclay which caused the layers to collapse. From XRD results, it can be inferred that nanoclays are well dispersed in EPDM-CIIR blends with intercalation, exfoliation and slight agglomeration [196].

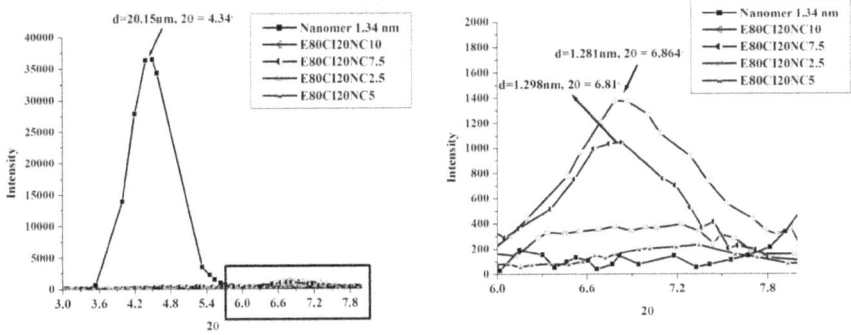

Figure 4.2 XRD spectra of EPDM-CIIR nanocomposites

TEM provides visual information on the morphology of nanoclay in the blend. For representation, TEM micrographs of the nanocomposites containing 5 phr and 10 phr nanoclay are shown in Figure 4.3(a) and (b) respectively. In nanocomposites containing lower phr of nanofiller, uniform dispersion of layered silicates with few stacks of clay platelets was observed. For E80CI20NC10, agglomerates or tactoids of silicates are visible. These agglomerates gave rise to the peaks observed in the XRD spectra.

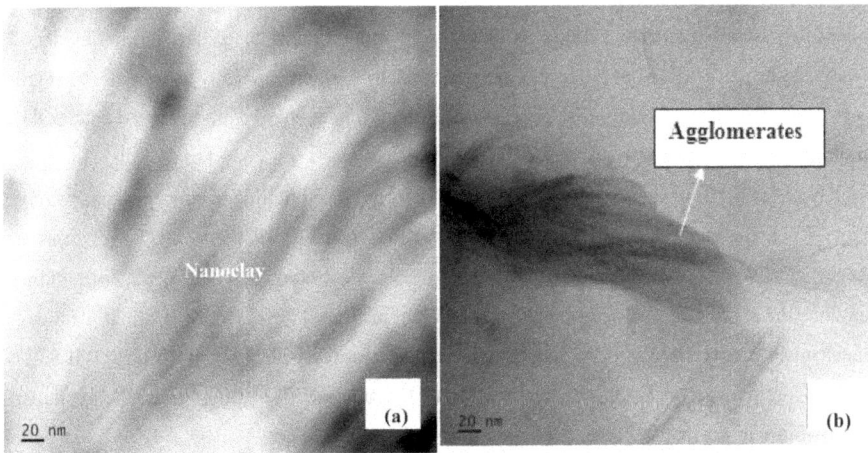

Figure 4.3 TEM micrographs of (a) E80CI20NC5 and (b) E80CI20NC10

The stacks of clay observed in TEM micrograph (E80CI20NC10) were responsible for the peaks in XRD spectra of nanocomposites containing higher content of nanoclay [197]. In conclusion, it was inferred from TEM and XRD that the organo-modified layered silicates were uniformly dispersed at lower filler loadings (5 phr) whereas slight aggregation was prevalent at higher nanofiller contents (10 phr).

4.2.4 Static mechanical properties

The static mechanical properties and stress-strain plot of nanoclay reinforced EPDM-CIIR blends are presented in Table 4.2 and Figure 4.4 respectively. The tensile strength (TS), modulus at 100 percent elongation (M100) and elongation at break (E_b) of the nanocomposites increased with increase in nanoclay content. The highest improvement in TS and M100 at 57% and 59% respectively was observed for E80CI20NC5. The evenly dispersed and exfoliated OMLS provided large interfacial area between the filler and elastomer chains. It was evident from morphological analysis that layered silicates had uniform exfoliation in E80CI20NC5. The polymer chains get entangled on the evenly dispersed and exfoliated layered silicate and are effectively reinforced [136]. The larger interface between the filler and the matrix facilitates better stress transfer to the filler. As a consequence, the nanocomposite can take up more loads, thus improving mechanical properties. Interface theories also provide explanation on how nanofillers improve mechanical properties. Upon incorporation of nanofiller, an interfacial zone is created at the interface of filler and the rubber matrix as schematically represented in Figure 4.5. The properties of interfacial zone are different from that of both filler and matrix. The mobility of rubber chains are restrained at the interface zone layer bound to the filler. In regions which are away from the vicinity of the interface zone, the mobility of rubber chains is more. Upto 5 phr, the increase in filler content gave rise to larger interfacial area and interactions and hence mechanical properties increased. At higher concentration of OMLS, the formation of agglomerates reduced the interfacial area and interactions [198]. Hence a slight decline in mechanical properties was observed for the nanocomposites with 7.5 and 10 phr OMLS. The reason for increase in E_b is due to the fact that organomodifier group in nanoclay imparts extensibility to the matrix [199].

Table 4.2 Static tensile properties of nanoclay reinforced EPDM-CIIR blends

Sample	Tensile strength MPa	Elongation at break E_b	M100 MPa	Percent increase in TS	Percent increase in M100
E80CI20	1.39± 0.07	95±8	0.92 ± 0.05		
E80CI20NC2.5	1.85±0.09	118±2	1.21 ± 0.04	33	28
E80CI20NC5	2.21±0.10	145±3	1.53 ± 0.02	59	57
E80CI20NC7.5	2.03±0.08	132±4	1.38 ± 0.03	46	43
E80CI20NC10	1.97±0.08	121 ± 3	1.24 ± 0.02	41	32

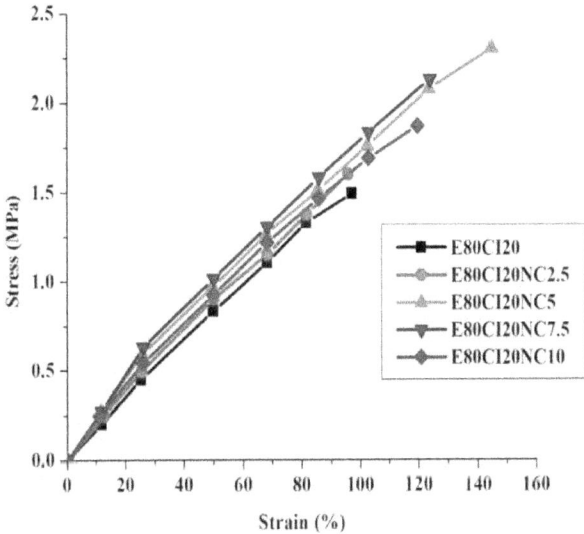

Figure 4.4 Stress strain response of nanoclay reinforced EPDM-CIIR blends

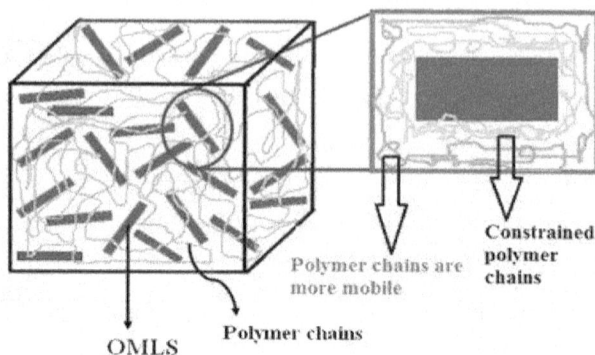

Figure 4.5 Schematic representation of nanoclay-rubber interface

4.2.5 Mooney Rivlin models

Mooney – Rivlin models are used to describe large scale non-linear behavior in elatomeric materials. The stress-strain plots were converted to Mooney-Rivlin curves, a plot of reduced stress (δ^*) vs $1/\lambda$. Reduced stress was calculated by by equation (4.1) [199].

$$\delta^* = \frac{\sigma}{\lambda - \lambda^{-2}} = 2\ (C_1 + C_2\lambda^{-1}) \tag{4.1.}$$

where σ is the applied stress, δ^* is the reduced stress, λ is the extension ratio and C_1 and C_2 are Mooney-Rivlin coefficients. Mooney-Rivlin curves were plotted with reduced stress δ^* vs λ^{-1} to analyze the reinforcing effect of nanofiller in the nanocomposites. The coefficients C_1 and C_2 are related to the filler network and flexibility of the elastomer chains, respectively [200], [201]. The Mooney-Rivlin plots for the nanocomposites are shown in Figure 4.6. It can be observed from Mooney-Rivlin plots that the reduced stress increased with filler content upto 5 phr. At higher nanoclay content, the reduced stress was lower than 5 phr, but higher than unfilled blends. As explained earlier, the reduction in δ^* at higher phr of nanoclay was due to agglomerate formation and lower interfacial area available for stress transfer [202]. From a different perspective, these plots can be regarded as V shaped curves. Due to lesser amount of filler content and rubber-filler interactions in E80CI20NC2.5 and E80CI20, the reduced stress is decreased with λ^{-1}. The filler networks are unstable and unable to take up more stress. In Mooney-Rivlin plot, V-shaped curve is seen from E80CI20NC5. The reduced stress increases at higher extension ratio (inverse) for nanocomposite with 5 phr and 7.5 phr nanoclay, due to more filler-rubber networks which is able to take up more stress.

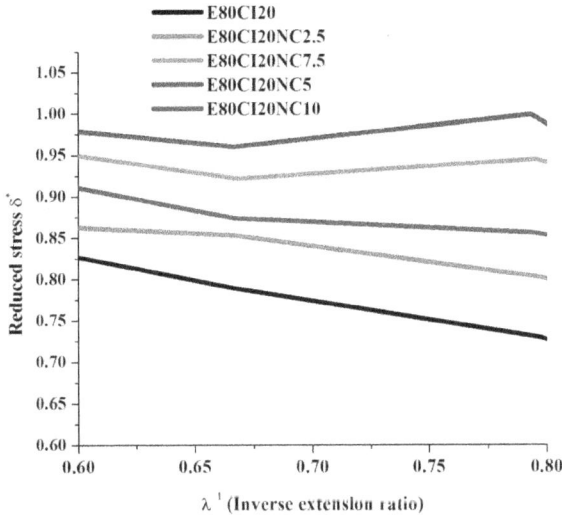

Figure 4.6 Mooney-Rivlin plots of nanoclay reinforced EPDM-CIIR blends

4.2.6 Theoretical modeling of nanoclay reinforced EPDM-CIIR blends

In this section, attempts have been made to examine the compliance of empirical and semi-empirical analytical models in predicting tensile modulus of the nanoclay reinforced nanocomposites [87] of EPDM-CIIR blends. The tensile modulus of polymer composites is dependent on several factors like matrix/ filler properties, content, alignment and aspect ratio of the filler . The models considered here are Voigt upper bound rule, Reuss lower bound rule, Hui-Shia, Guth, Modified Guth, Halpin-Tsai and modified Halpin-Tsai models. The equations corresponding to these models are summarized in Table 4.3. Experimentally obtained elastic moduli of nanoclay reinforced EPDM-CIIR elastomer blends were compared with theoretical predictions with the models and are presented in Table 4.4.

Table 4.3 Micromechanical models for predicting moduli of EPDM –CIIR nanocomposites

Model	Equation
Voigt upper bound rule (Rule of mixtures)	$E = \Phi_p E_p + (1-\Phi_p)E_m$
Reuss lower bound rule (Inverse rule of mixtures)	$\dfrac{1}{E} = \dfrac{\Phi p}{Ep} + \dfrac{(1-\Phi p)}{Em}$
Guth model	$E = E_m\,[1+2.5\Phi_p + 14.1\ \Phi_p{}^2]$
Modified-Guth model	$E = E_m[1 + 0.67\ \alpha\Phi_p + 1.62\ (\alpha\Phi_p)^2]$
Halpin – Tsai	$E = E_m\,[\dfrac{(1+\xi\eta\Phi p)}{(1-\xi\eta\Phi p)}]$ $\eta = \dfrac{[(\frac{Ep}{Em}) - 1]}{[(\frac{Ep}{Em}) +\xi\,]}$
Modified Halpin- Tsai equations	$E = E_m\,[\dfrac{(1+2\,(\frac{w}{t})\,\Phi p)}{(1-2\,(\frac{w}{t})\Phi p)}]$ Since $E_p >>> E_m$, therefore $[(\frac{Ep}{Em}) - 1] \cong [\ (\frac{Ep}{Em}) +\xi\ ]$
Hui- Shia model	$E = E_m\,\{1- (\dfrac{\Phi p}{4})\ [(\frac{1}{\xi})+ (\frac{3}{\xi})+\wedge]\}^{-1}$ where $\xi = \Phi_p + [(\dfrac{E_m}{E_p - E_m})] + 3(1-\Phi_p)\,\{\dfrac{[(1-g)\alpha^2 - (g/2)]}{(\alpha^2-1)}\}$ $g=(\dfrac{\pi\alpha}{2})$ and $\wedge= (1-\Phi_p)\,\{\dfrac{[3(\alpha^2 + 0.25)g - 2\alpha^2]}{(\alpha^2-1)}\}$

Parameters

E Moduli of composite, E_p Moduli of filler, E_m Moduli of matrix

Φ_p Volume fraction of the filler

α Shape factor ($\alpha = \dfrac{length}{breadth}$) accounted for "accelerated stiffening". α is the inverse aspect ratio

v Poisson's ratio (= 0.5) assumed to be same for both the components and perfect interfacial interactions between filler and matrix

ξ Shape parameter, which is a function of geometry of filler and the alignment in the matrix, ξ for layered silicate type filler is $2(\frac{w}{t})$, w and t are width and thickness of the exfoliated filler

The experimental elastic modulus of E80CI20NC5 was in close compliance with that predicted by Hui-Shia model. The Hui-Shia model considered was developed with assumptions like (i) perfect interaction between filler and matrix (ii) uniformity in size, geometry, and orientation of fillers (iii) aligned reinforcements with fibre or flake like fillers and (iv) matrix and filler are linearly elastic and isotropic [203].These conditions were largely fulfilled by the nanocomposite with 5 phr organo-modified nanoclay. The nanofiller in E80CI20NC5 were well dispersed with good interfacial interaction between the filler and rubber as evident from morphological and mechanical analysis. But for higher filler contents (for 7.5 and 10 phr nanoclay), Reuss model would be appropriate. The agglomerates of layered silicates in these blends behaved like microfillers [119]. The properties of the nanocomposites are largely governed by physical and chemical interactions at the rubber-nanofiller interface [97]. Many of the micromechanical models donot consider these interactions and hence the properties predicted by those models deviate from observed values [204].

Table 4.4 Comparison of tensile moduli from experimental observation with theoretical prediction

content (phr)	E (exp)	Voigt Upper Bound rule	Reuss lower bound rule	Hui-Shia	Halpin-Tsai	Guth	Modified Guth	% Diff Reuss	% Diff Hui-Shia	% Diff Guth
0	0.98	0.98	0.98	0.98	0.98	0.98	0.980	0	0	0
2.5	1.25	15846	1.11	1.41	50.9	1.40	829	11.6	-12.7	-11.9
5	1.55	28470	1.23	1.54	100.9	1.73	2653	20.6	0.01	-12.2
7.5	1.41	38762	1.36	1.64	150.9	2.04	4905	3.54	-16.3	-44.7
10	1.26	47316	1.48	1.73	200.7	2.23	7299	-18.4	-37.3	-76.9

4.2.7 Rheological characteristics of nanoclay reinforced EPDM-CIIR elastomeric blends

Mooney viscosity, Delta Mooney (the difference between initial value (IV) and final value (FV)), scorch time (t_5), vulcanization time (t_{35}) and vulcanization index ($\Delta t = t_{35}\text{-}t_5$) of the nanocomposites were evaluated and summarized in Table 4.5. Scorch time denotes the time during which a rubber can be worked at a predetermined temperature before curing starts [202]. The decrease in t_5 with nanofiller reinforcement in matrix can be attributed to higher shear heating during mixing [205]–[207]. High shear heating tends to induce premature vulcanization that produces higher compound viscosities. t_{35} increased as a function of OMLS content owing to restrictions imposed by nanoclay on the molecular mobility of rubber chains. The increase in curing or crosslinking between rubber chains for nanocomposite can be explained by a different phenomenon. The viscosity of the nanocomposite gets enhanced when the mobility of rubber chains gets restricted after reinforcement [207]. In addition, the nanofiller impose additional resistance to the formation of crosslinks in between rubber chains. Hence more torque is required for vulcanization or curing reaction of stiffer nanocomposites, which increases the curing time. The cure characteristics from moving die rheometer (MDR) or mooney viscometer are tabulated in Table 4.5. Mooney viscosity can be defined as the shearing torque resisting rotation of cylindrical disk enclosed in a chamber. Percentage change in Mooney viscosity ($ML_{(1+4)}$) for OMLS incorporated blends was found to be 142 % or more than neat blends. The decline in mooney value at higher filler contents (10 phr) was because of formation of agglomerates that increased the mobility of the rubber chains making the compound less stiffer [208].

Table 4.5 Mooney viscosity characteristics of nanoclay reinforced EPDM-CIIR blends

MDR characteristics	E80CI20	E80CI20NC2.5	E80CI20NC5	E80CI20NC7.5	E80CI20NC10
IV (mooney)	56.5	102	107	121	55.3
FV (mooney)	25.6	27.3	29.5	32.9	32.4
Delta mooney (IV-MV)	30.9	74.7	77.3	88.3	22.9
% change in mooney viscosity	-	142	150	186	25.8
t_5 (minutes)	8.35	7.65	6.82	6.15	6.35

t_{35} (minutes)	11.86	12.3	12.53	12.73	14.1
Optimum cure time t_{90} (minutes)	7.15	7.27	7.42	8.27	8.28

4.2.8 Dynamic Mechanical Behavior

Dynamic Mechanical Analysis (DMA) is an effective method to investigate the chain mobility as well as the behavior of the material under various conditions of stress, temperature etc. The viscoelastic properties like storage modulus (E'), loss modulus (E") and tanδ (E"/E') are obtained from DMA when the material is subjected to sinusoidal deformation as a function of temperature or frequency. Marcos and Hatsuo [209] studied dynamic mechanical response of viscous and elastomeric silicone composite of polyimide material. The presences of interface layer produced by silicone coating were observed from a new relaxation peak [209]. Shyan and Hatsuo [210] utilized dynamic mechanical tests to determine glass transition temperature, crosslink density and activation energy for glass-transition process for high performance polymers. They correlated degree of cure with increase of storage and loss modulus after glass transition. The height of tanδ peak associated at transition range were related to degree of cure and crosslink density [210]. DMA was also used to describe both quantitative and qualitative behavior of rubber/epoxy blends [211].

(a) Storage Modulus (E')

The storage (elastic) modulus (E') for unfilled EPDM-CIIR blend and nanoclay reinforced EPDM-CIIR blends are plotted in Figure 4.7. The E' value provides information about the stiffness of the material, a measure of elastic nature of the material. The dynamic elastic modulus (E') increased noticeably with the increase in filler levels in the temperature range -70°C to -30°C. Enhancement in storage modulus (E') with increase in filler content was because of higher filler-polymer interactions [212], [213]. In nanoclay reinforced blend, the high aspect ratio of nanoclay and the formation of intercalated and exfoliated structures result in higher degree of filler – polymer interactions than micron sized fillers. The value of E' was maximum for blend reinforced with 5 phr of OMLS indicating superior "reinforcing effect" [169], [197], [199]. The

82

curves representing storage modulus show three distinct regions representing glassy regions with high modulus where the mobility of rubber chains are restricted, transition region where considerable decrease in modulus can be observed with increase in temperature and a rubbery region where modulus remains constant over an increase of temperature [214]. The storage modulus of nanocomposites in the rubbery plateau region (between 0°C and 30°C) was also relatively higher [211] than neat blend. It was also found that the crosslink density of EPDM-CIIR nanocomposites, as determined from solvent sorption studies discussed in next section of this chapter were higher than unfilled EPDM-CIIR blends. The increase in crosslink density also restricted conformational freedom and movement of EPDM-CIIR chains around nano-reinforcements [51].

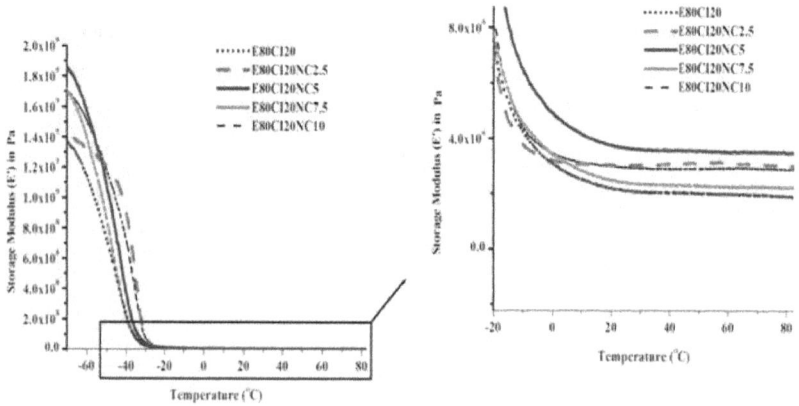

Figure 4.7 Storage (elastic) modulus of nanoclay reinforced EPDM-CIIR blends

(b) Reinforcing efficiency (r)

The efficiency factor 'r' for reinforcement mechanism is calculated from the following equation

$$E_c = E_m (1 + rV_f) \qquad (4.2.)$$

The dispersion mechanism (filler-filler and filler-polymer networks) and interfacial interactions (hydrodynamic reinforcement effect) influences the reinforcement efficiency. From Table 4.6, it is evident that factor of reinforcement efficiency increases with organo-modified layered silicate

(OMLS) or nanoclay content up till 5 phr. The increase in reinforcing efficiency with OMLS content is attributed to ability of nanofillers to take up applied stress. The mechanism of reinforcement is explained in the previous sections.

(c) Loss Modulus (E")

The variation of loss modulus (E") as a function of temperature for varying layered silicate levels are plotted in Figure 4.8. The E" is the viscous modulus associated with dissipation of energy. E" decreased with increase in filler levels. The loss occurs when polymer chains slide past each other or the filler. The entanglement of polymer chains around the filler and reinforcement by the nanofiller prevents the chains from slipping past each other. Hence E" decrease with nanoclay content. However at 10 phr nanofiller content, formation of agglomerates reduces the reinforcement and the chains slip past each other more easily than in nanocomposites with 5 phr nanoclay content. Hence E" in E80CI20NC10 is higher than E80CI20NC5 [94], [95].

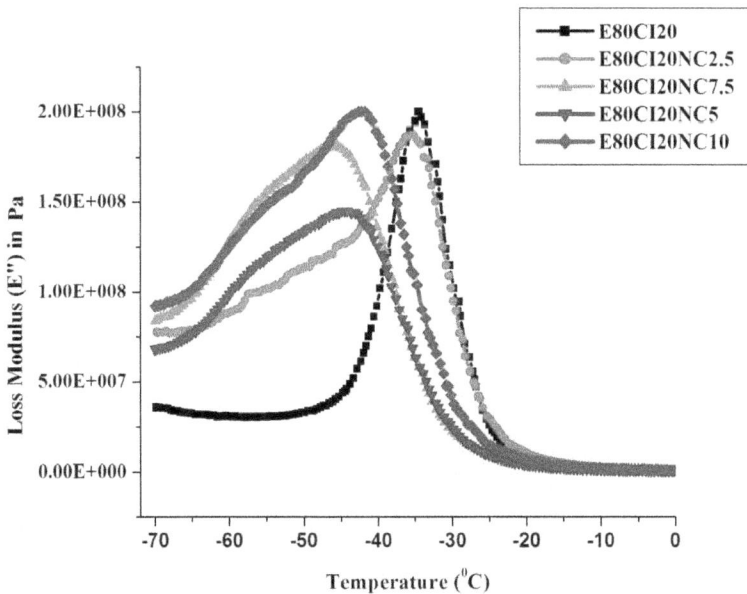

Figure 4.8 Loss modulus (viscous) of nanoclay reinforced EPDM-CIIR blends

(d) Effectiveness of filler (C)

The effectiveness coefficient (C) of moduli of the nanocomposite is calculated from the ratio of storage modulus in glassy to rubbery regions of the nanocomposite to that of unfilled blend, as given by equation 4.3.

$$C = \frac{(E'\text{Glassy}/\text{rubbery})\text{nanocomposite}}{(E'\text{Glassy}/E'\text{Rubbery})\text{blend}} \qquad (4.3.)$$

where E'_{Glassy} and E'_{Rubbery} are the storage modulus at glassy state (-70°C) and rubbery state (70°C) respectively. From the values of filler effectiveness coefficient C, tabulated in Table 4.6, it can be inferred that maximum stress transfer between filler and matrix occurred at E80CI20NC5 and thereafter effectiveness decreased.

Table 4.6 Reinforcing parameters obtained from DMA analysis of EPDM-CIIR nanocomposites.

Sample	Reinforcement factor	Measure of entanglement (* 10^{-4} moles/m^3)	Volume fraction of constrained region C_v
E80CI20NC2.5	2.43	0.12	0.051
E80CI20NC5	2.75	0.14	0.425
E80CI20NC7.5	1.47	0.11	0.399
E80CI20NC10	0.87	0.09	0.375

(e) Loss tangent (tanδ)

The loss tangent (tanδ) of EPDM-CIIR nanocomposites as a function of temperature at 1Hz frequency is plotted in Figure 4.9. The reinforcement of the blend with nanoclay reduced the damping properties of the elastomer. Loss tangent values provide information about relaxation behavior of nanocomposites, as well as the energy dissipation due to internal friction and molecular motions. The lowering of loss tangent peak was seen for nanocomposites after 2.5 phr nanoclay or OMLS content. The significant lowering and broadening of peak observed at 5 and

7.5 phr OMLS was because of constrained segmental movements of polymer chains in the matrix. When temperature increases, the molecular mobility is facilitated due to thermal agitation. At the transition zone between glassy and rubbery region, elastic modulus decreased monotonically, loss modulus and tanδ pass through a maximum [67], [209]. The reduction in T_g (glass transition temperature) in the nanocomposites can be attributed to increase in extensibility provided by organo-modifier group in nanoclay [90]. The exfoliated platelets of layered silciates separate polymer chains and reduce chain to chain interaction also reduces T_g. The reinforcement, rubber-OMLS interactions and constrained segmental mobility of the rubber chains reduce the relaxation amplitude of the rubber chain segments. Hence, the loss tangent peak lowers on addition of OMLS to the blends. The lowering of loss tangent is most significant in blends containing OMLS 5 phr and 7.5 phr. The lower peak implies that the amount of energy dissipated by nanocomposites is lower than that dissipated by unfilled blends.

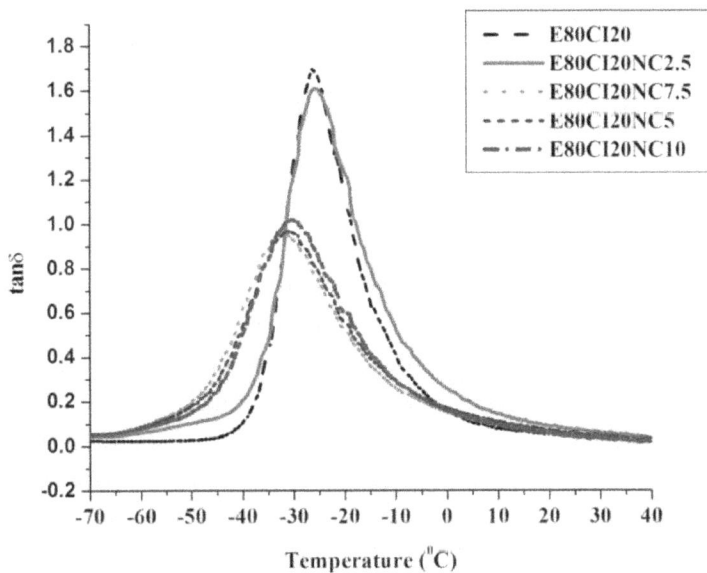

Figure 4.9 Tan (delta) values of nanoclay reinforced EPDM-CIIR blends

Table 4.7 Dynamic mechanical properties of nanoclay reinforced EPDM-CIIR blends

Sample	$\tan\delta_{max}$	E"$_{max}$ (GPa)	T_g from $\tan\delta$*(°C)	T_g from E" (°C)	C (10Hz)	ΔG' (1%-300% strain range)
E80CI20	1.66	2.00	-26.4	-34.4	1	229
E80CI20NC2.5	1.61	1.87	-26.2	-35.5	0.15	321
E80CI20NC5	0.95	1.45	-31.4	-44.2	0.58	341
E80CI20NC7.5	0.96	1.81	-31.1	-45.5	0.19	333
E80CI20NC10	1.02	2.00	-30.5	-41.8	0.21	277

(f) Constrained volume (C$_v$)

Generally in elastomers, nano reinforcements lead to formation of an immobile polymer chains near the filler surface in nano layer as shown in Figure 4.5. The glass transition of a rubber is influenced by various factors such as constraints associated at amorphous phase by crosslinking, free volume, molecular packing, etc. The constrained chains are estimated based on the reduction in number of mobile polymer chains during glass transition as represented in schematic diagram (Figure 4.5). Broadening of loss tangent peak is also due to the restriction in slippage of rubber chains near constrained regions near OMLS-rubber interfaces. These regions are formed by filler-polymer networks and reinforcing effect. High thermal agitation is required to initiate the slippage of entangled rubber chains [210] near OMLS in nanocomposites. The prediction of volume fraction of constrained region is essential in order to explain the enhancement in mechanical properties with nanoreinforcement. The volume fraction of constrained region C$_v$ can be estimated from energy loss fraction W of the polymer nanocomposite.

$$W = \frac{\pi\tan\delta}{\pi\tan\delta + 1} \qquad (4.4.)$$

The W at the loss tangent (tanδ) peak is given by the following relation

$$W = \frac{(1-C_v)\,W_o}{(1-C_o)} \qquad\qquad (4.5.)$$

where $(1- C_v)$ is the fraction of amorphous region, C_o and W_o represents volume fraction of constrained region and energy fraction loss for neat EPDM-CIIR blend respectively. The volume fraction of constrained region C_v increases with OMLS content as given in Table 4.6. The effective constrained volume is maximum for E80CI20NC5 due to its extended constraint zone near filler interfaces, indicating better reinforcing effect.

(g) Cole-Cole plot

The structural changes taking place after incorporation of nanoclay to EPDM-CIIR blends can be studied using Cole-Cole plot, where loss modulus (E") were plotted with storage modulus (E') at a particular frequency. The Cole-Cole plot also known as Wicket plot was effectively used to identify the homogeneity or heterogeneity in nanocomposites based on nature of the plot. Homogeneity of the nanocomposite system was indicated by smooth, semicircular arc in the Cole-Cole plot while irregular or imperfect shape in the plot signifies heterogeneity in the nanocomposite. The plot of storage modulus E' vs loss modulus E" for OMLS reinforced EPDM-CIIR blends was shown in Figure 4 10 (a). For 2.5 and 5 phr nanoclay reinforced blends, the curve is superimposed with neat blend indicating heterogeneity in dispersion of layered silicates. The irregular shape in Cole-Cole plot seen for E80CI20NC7.5 and E80CI20NC10 confirms heterogeneity to the structure of the nanocomposite. From the modified Cole plot (log E' Vs log E") in Figure 4.10(b), it can be observed that the plot for nanocomposites was superimposed with that of the unfilled blends implying that there was no variation in the structure of the elastomer system upon incorporation of filler.

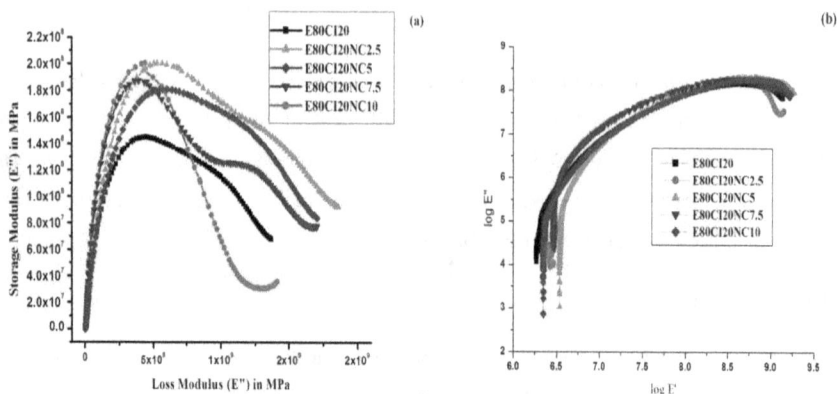

Figure 4.10 (a) Cole-cole plot and (b) modified cole plot of EPDM-CIIR nanocomposites

(h) Analytical models for predicting dynamic elastic (storage) modulus

Several analytical models like Einstein, modified Einstein, Liang, Halpin-Tsai and Quemada equations that correlate storage modulus with filler content are available in the literature [114], [215], [216]. These models are summarized in Table 4.8. The relation between storage modulus of filled and unfilled elastomer are characterized by the relative storage modulus (E_R'). E_R' is the ratio of storage modulus of OMLS reinforced ones to that of unfilled blends. Liang developed a model after taking into consideration the inclusions and their interactions with the matrix. The experimental E_R' and E_R' predicted by the various models are plotted in Figure 4.11 for different nanoclay content in the nanocomposite.

The experimental value of E_R' of E80CI20NC5 showed good agreement with those predicted by Liang model. The reason for compliance of experimental and predicted value as with the case of static modulus is well dispersed filler and good filler-polymer interaction in E80CI20NC5.

Table 4.8 Micromechanical models for predicting dynamic storage modulus

Model	Equation
Einstein equation	$E_R{}' = 1 + 2.5\Phi_f$
Guth	$E_R{}' = 1 + 2.5\Phi_f + 14.1\Phi_f^2$, modified Einstein equation by adding particle interaction term
Halpin and Tsai equation	$E_R{}' = \dfrac{(1 + \tau\eta\Phi_f)}{(1 - \eta\Phi_f)}$
Liang model	$E_R{}' = 1 + \dfrac{\tau\Phi_f(m'-1)}{1 + (1-\Phi_f)(m'-1)\lambda}$ and $\lambda = \dfrac{7 - 5v_m}{15(1 - v_m)}$ When the storage modulus of nanoclay is higher than the matrix, the above equation gets simplified to $E_R{}' = 1 + \dfrac{\tau\Phi f}{(1 - \Phi f)\lambda}$

Parameters

Φ_f Filler volume fraction

τ Degree of reinforcement, depends of filler geometry and loading conditions

$\eta = \dfrac{m-1}{m+\tau}$ and m= $\dfrac{E_f}{E_m}$ where E_f and E_m are modulus of OMLS and neat matrix respectively

v_m Poisson ratio (=0.5)

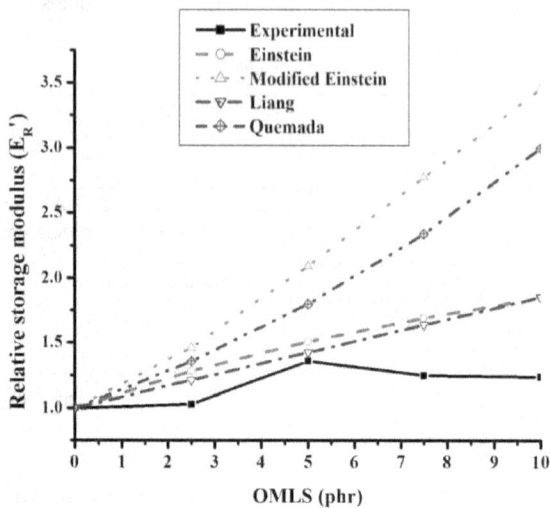

Figure 4.11 Variation of relative storage modulus with various analytical models

4.2.9 Payne Effect

Payne effect describes the phenomenon that storage modulus (G') of the nanofiller reinforced elastomer decreases with increasing applied dynamic strain [149], [209]. The amplitude of Payne effect ($\Delta G = G_0' - G_\infty'$) is dependent on the structural and surface characteristics, specific surface area and distribution of filler in the matrix. Here G_0' is taken at 1% dynamic strain and G_∞' at 300% strain. Increase in ΔG with increase in OMLS content, as tabulated in Table 4.7, proved that Payne effect was significant in the nanocomposites.

The viscoelastic behaviour of OMLS reinforced EPDM-CIIR blends are dependent on the interactions taking place within the polymer nanocomposites viz. filler-filler and filler-polymer interactions [97]. Filler-matrix interactions produced from nano reinforcements results in high storage modulus at lower strains. The dynamic storage moduli (G') are plotted as a function of dynamic strains in Figure 4.12 (a). The G' was highest for E80CI20NC5 similar to trends observed in static and dynamic mechanical properties. The storage modulus of filled rubber deviates from a plateau value (G_0') and collapses to a minimum value (G_∞') beyond a particular dynamic strain. The decline in G' was uniform and gradual for 5 phr OMLS reinforced blend. This behavior was ascribed to better interfacial filler-rubber interactions as discussed in earlier sections. In the case of E80CI20NC7.5 and E80CI20NC10 the drop in storage modulus was sudden due to rupture of filler-filler networks. A schematic representation depicting the difference in Payne effect in nanocomposites with well dispersed and agglomerated OMLS is given in Figure 4.13 (a) and (b) respectively.

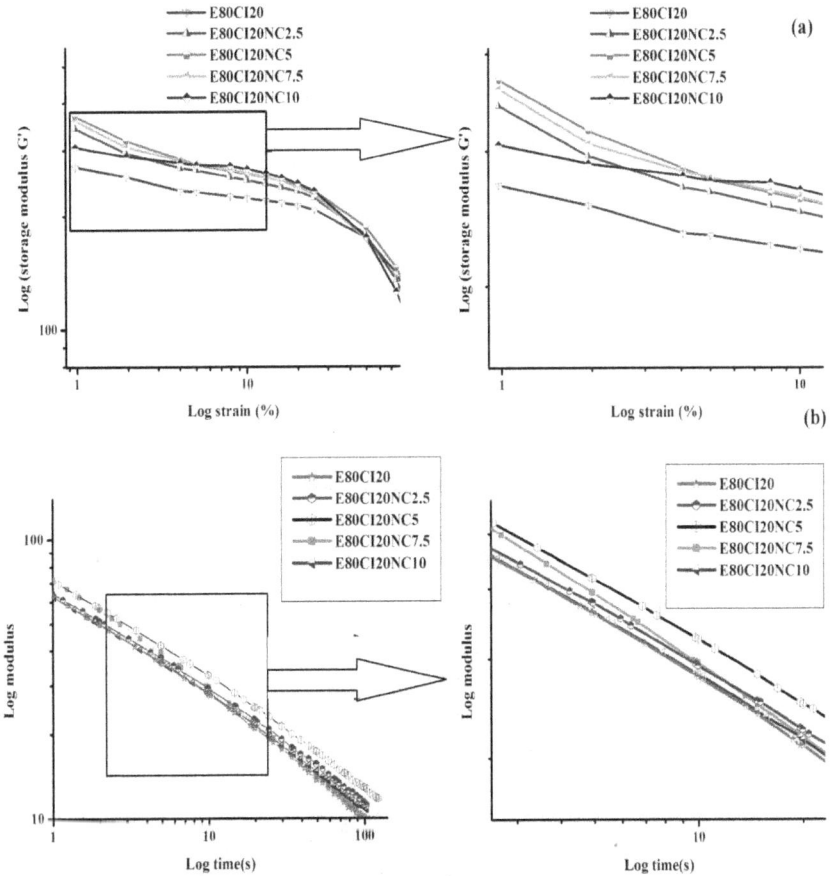

Figure 4.12 (a) Payne effect and (b) stress relaxation of nanoclay reinforced EPDM-CIIR blends

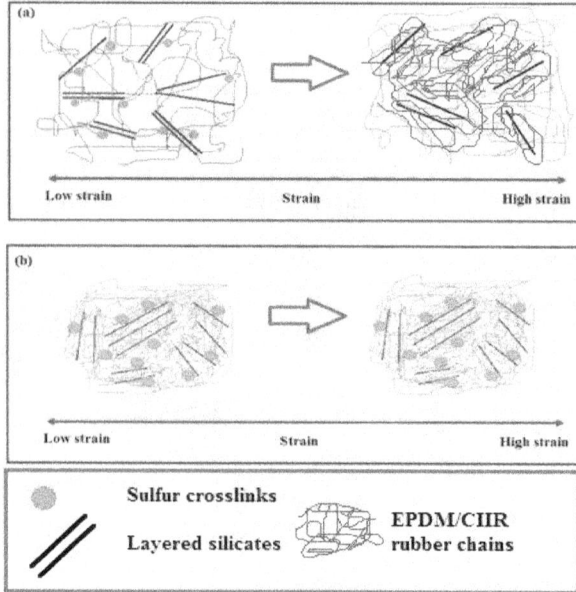

Figure 4.13 Schematic representations of (a) filler-rubber and (b) filler-filler interactions

The physical interpretation of the influence of dynamic strain on elastic modulus was first proposed by Payne [217]. Comparison of experimental data with non-linear viscoelastic models can provide more inter retations on Payne effect. Kraus proposed a model considering the agglomeration/de-agglomeration effects of filler aggregates by assuming Vander Waal's type of interactive forces between filler particles [218]. The applicability of Kraus model to the EPDM-CIIR nanocomposites was studied. The Kraus model given in the following equation is based on agglomeration/de-agglomeration effect of filler aggregates

$$\frac{G'(\gamma)-G'(\infty)}{G'_0-G'(\infty)} = \frac{1}{1+(\frac{\gamma_0}{\gamma_c})^{2m}} \qquad (4.6.)$$

where $G'(\gamma)$ is the dynamic shear modulus obtained at particular strain (γ), $G'(\infty)$ is the value of storage modulus at very large strain, γ_c is the critical strain (ie. particular strain at which $G'_0 - G'(\infty)$ is half of its initial low strain value) and m is the shear strain sensitivity which indicates the shape of the strain curves. The dynamic shear modulus at particular strain obtained from

Kraus model did not fit with experimental data for EPDM-CIIR nanocomposites. According to literature reports, the Kraus model provides fitting with experimental data for rubbers with larger filler aggregates [123], [219], [220]. Several researchers have reported that Kraus model was in close tolerance for viscoelastic behavior of larger sized carbon black or micro silica filled rubbers. In case of nanocomposites, larger interfacial area created by specific surface area of the filler and small particle size attributes towards formation of smaller tactoids of nanofiller.

(a) Loss factor under dynamic strain sweep conditions

The tanδ (loss factor) of OMLS reinforced EPDM blends under dynamic strain sweep conditions are plotted in Figure 4.14. Since there was no noticeable change in tanδ at lower strains, tanδ was plotted for dynamic strain range from 60% to 240%. tanδ increased with increase in strain for all nanocomposites. tanδ decreased with increase in layered silicate content, which was consistent with that observed in DMA analysis. This behavior of tanδ indicated that internal frictional loss in nanocomposites was lower than unreinforced blends. Good dispersion of layered silicate in the elastomer obstructed the rubber molecules from slipping on the surface of fillers [221]. Thus energy loss and intrinsic friction in nanocomposites reduced drastically, a characteristic that is required for continuous loads bearing applications of rubber components.

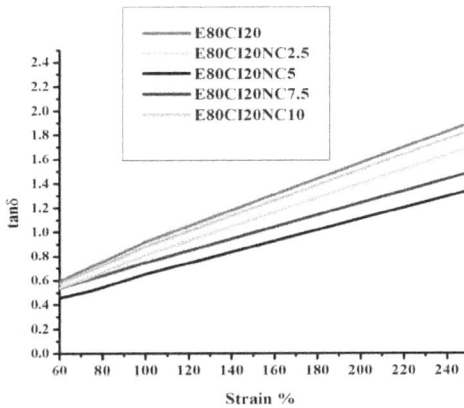

Figure 4.14 Tan(delta) for OMLS reinforced EPDM blends under strain sweep conditions

4.2.10 Stress relaxation test (or force decay) analysis

Stress relaxation or force decay analysis was performed to analyze the variation in the decline of stress with time as a function of nanofiller level [136]. These tests were performed on specimens sheared to a particular strain or position. The decline in stress/modulus with time for nanocomposites is shown in Figure 4.12 (b). The modulus decay was due to the chemical and physical changes incurred when the material is strained. The time-dependent physical relaxation in rubber takes place due to rearrangement of molecular chains and fillers. It was seen that a sudden decay in modulus occurred at higher filler content (10 phr) due to breaking down of filler tactoids. A slow and uniform stress relaxation or force decay was observed for nanocomposite with exfoliated morphology (5 phr). It is important to note that the relaxation occurred in two phases because of different types of interactions present in elastomer nanocomposites. The initial phase relates to breakage of filler-filler networks and final phase indicated a rupture of jammed filler-polymer networks.

4.2.11 Solvent sorption behavior

The solvent sorption characteristics of an elastomer nanocomposites is a function of several factors like nature of elastomer, polymer segmental mobility, nanofiller geometry and concentration, filler orientation and dispersion in the rubber, nature of solvent, size of the penetrant molecules, temperature, possible reaction between solvent and the nanocomposite etc. The sorption characteristics of nanocomposites are effective tools to understand the morphology and polymer –filler interaction in the nanocomposites. The sorption behavior of EPDM – CIIR nanoclay composites were evaluated by analyzing the sorption curves represented in Figure 4.15, a plot of moles of solvent uptake per 100 g of rubber (Q_t) as a function of √t (time).

The sorption and equilibrium sorption for the nanocomposites were considerably lower when compared with unfilled blend. The permeability of a solvent through an elastomer is a direct function of the concentration of space available in the matrix that is large enough to hold the penetrant molecule. Incorporation of nanoclay in the elastomer matrix led to reduction in availability of these spaces with simultaneous restriction in elastomer segmental mobility and creation of "tortuous path" for solvent molecule transport. The solvent uptake was reduced with the increase in nanoclay content, which implied that the barrier property was enhanced. In order to assess the swelling behavior of the nanoclay reinforced EPDM CIIR blends, the swelling coefficient (β) was calculated by the equation 3.8 given in previous chapter [222]. In comparison with unfilled E80CI20, the swelling coefficient decreased with the increase in nanoclay content, as tabulated in Table 4.9.

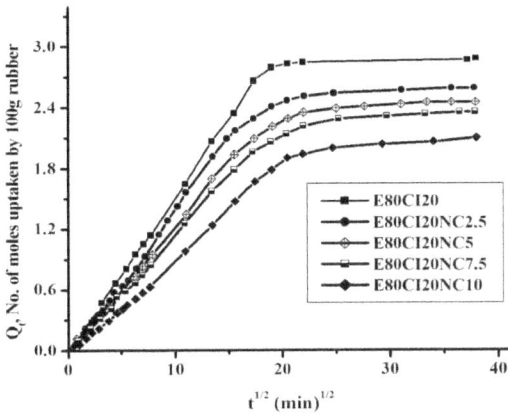

Figure 4.15 Sorption isotherms of nanoclay reinforced EPDM-CIIR blends

With increase in the nanofiller content, the availability of free volume in the matrix and the ease with which the polymer chain segments exchange their position with the penetrant molecules decreased and hence diffusion coefficient, calculated with equation 3.12 also reduced. The sorption coefficient, which is the ratio of the mass of solvent uptake at equilibrium to initial mass of specimen, is calculated from the sorption isotherms. The permeation coefficient or permeability which is the product of diffusion and sorption coefficient (P=D*S) is also tabulated. From the Table 4.9, it can be inferred that the transport coefficients decreased with nanoclay content. Platelet like morphology of layered silicate and its uniform dispersion in the blend produced a tortuous pathway for the penetrant to traverse through the nanocomposite. This also resulted in reduction in available free volume in the blend matrix [88], [223], [224]. Nanoclay filled blends had lower M_c values compared with E80CI20. Hence the amount of equilibrium uptake and coefficients of diffusion, sorption and permeation decreased with increase in nanoclay content [196].

Table 4.9 Transport properties of EPDM-CIIR nanocomposites

Sample	Diffusion coefficient $(Dx10^7)$ m^2/s	Sorption coefficient (S) (g/g)	Permeability coefficient $(P \times 10^7)$ (m^2/s)	Swelling coefficient (β) (cm^3/g)	Molar mass between cross-links $(cm^3/gmol)$	Cross link density x $10^4(gmol/cm$
E80CI20	3.51	1.59	5.58	2.04	2468	4.05
E80CI20NC2.5	3.24	1.52	4.92	1.95	2128	4.69
E80CI20NC5	2.81	1.35	3.79	1.73	1868	5.35
E80CI20NC7.5	2.64	1.03	2.72	1.32	1732	5.77
E80CI20NC10	2.15	0.76	1.65	0.97	1125	8.89

4.2.12 Thermogravimetric analysis

The thermal stability and the extent of thermal degradation of the specimens were determined using TGA. The results of TGA and derivative thermogravimetry (DTG) are shown in Figures 4.16 (a) and (b) respectively.

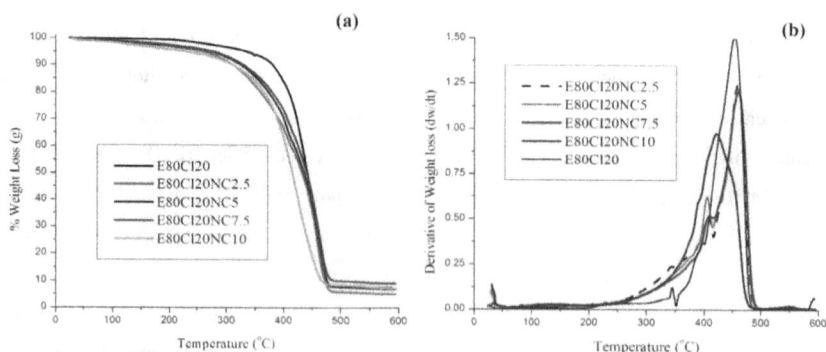

Figure 4.16 (a) TGA and (b) DTG thermograms of EPDM-CIIR nanocomposites

The peak degradation temperature (the temperature at which maximum degradation occurs) increased from 450°C in E80CI20 to 460°C in E80CI20NC5. Dispersion of layered silicate in a polymer matrix forming dispersed or exfoliated nanostructures restricted thermal motion of polymer chains and may be responsible for the rise in peak degradation temperature Generally, thermal stability enhanced on incorporation of nanoclay [48], [133].

4.2.13 γ- Irradiation studies of nanoclay reinforced EPDM-CIIR blends

The EPDM-CIIR nanoclay composites were exposed to ^{60}Co-γ radiation and exposure to γ-radiation resulted in cross-linking and/or chain scission, depending on the radiation dose. Consequently, the mechanical properties, transport coefficients and thermal degradation of the nanocomposites were altered depending on radiation doses and nanoclay content [225].

4.2.13 (a) FTIR spectra of irradiated EPDM-CIIR nanocomposites

To understand the chemical changes in the nanocomposites on exposure to gamma radiation, FTIR spectra of irradiated specimens were compared with unirradiated ones [184]. FTIR spectra were analyzed for all irradiated nanocomposites. As a representation, the comparison for E80CI20NC5 is presented in Figure 4.17. The spectrum of unirradiated E80CI20NC5 showed intense C-H symmetric and asymmetric stretching peaks located at 2850cm^{-1} and 2919 cm^{-1} that corresponds to the main chain methylene group (-CH$_2$) [226][227]. Absorption peaks were seen below 1500 cm^{-1} due to various C-H deformations. For unirradiated E80CI20NC5, the C-O-C stretching vibration of methyl carbonyl group was found at 1147cm^{-1} [228]. The absorption peak found at 1247cm^{-1} arising from C-C and C-O bonds stretching in un-irradiated specimens gave rise to C-O-H bending in irradiated nanocomposites [16]. For 0.5 and 1MGy doses of gamma radiation, new peaks are found at 1390 cm^{-1} corresponding to symmetric stretching of carboxylate (COO$^-$) ions and those near to 1230cm^{-1} arise from C-O stretching, in addition to C-C bonds corresponding to 1642 cm^{-1}. The noticeable peak at 1647 cm^{-1} can be attributed to the stretching C=C group which increased with intensity of radiation till 1 MGy indicating radiation induced crosslinking. No significant peaks were observed for nanocomposite irradiated at 0.5 MGy. Gamma irradiation of polymers resulted in bond cleavage giving rise to free radicals, which in presence of oxygen react to form hydroxyl group by chain mechanism [229]. The free radicals produced by gamma irradiation react with oxygen molecules to form peroxy radicals, which can remove the hydrogen atom from polymer chain to form hydroperoxide [230]–[232]. The hydroperoxide further splits into two new free radicals viz. hydroxyl and peroxy radical that continue to propagate and cause further degradation [233]. Studies have shown that the damage caused to elastomeric components subjected to radiation are cross linking and chain scission, chemical interaction with environmental agents, especially oxygen and both radiation induced crosslinking and degradation process takes place simultaneously but at different rates [227], [228]. An intense new band was found at 3337.01 cm^{-1} for nanocomposite subjected to highest dose of gamma irradiation that can be assigned to stretching of –OH group as a result of free radical generation by chain scission [40], [230], [234], [235].

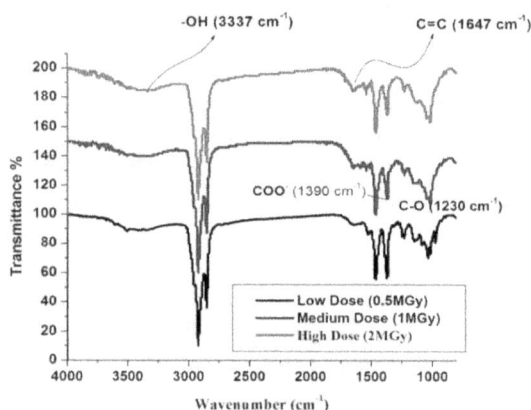

Figure 4.17 FTIR spectra of irradiated nanocomposite (E80CI20NC5)

4.2.13 (b) Mechanical properties after gamma-irradiation

The mechanical properties of irradiated EPDM-CIIR nanocomposites are tabulated in Table 4.10 and Figure 4.18. It is evident that the tensile strength and modulus of the nanocomposites were found to increase after irradiation for radiation doses of 0.5 and 1 MGy. This indicated that the dominant reaction was crosslinking, as evident from FTIR analysis. However at 2MGy dose, chain scission prevailed resulting in reduction in mechanical properties. At all cumulative doses of radiation, E80CI20NC5 showed minimum change in properties from unirradiated samples, as listed in Table 4.10. This is due to the uniform dispersion and exfoliation of nanoclay in

E80CI20NC5 that provided tortuous path and interfaces for the free radical transport. Further, the antioxidant characteristics of the organo-modified MMT (nanoclay) aided in absorption of the free radicals generated during exposure to radiation, thereby delaying degradation and enhancing thermal stability [45], [236]. Crosslinking generates chemical bonds between adjacent polymer molecules and thereby the stiffness rise. The functional group of layered silicates also may find additional sites for cross-links. The organo-modified nanoclays are effective radiation damage inhibitors in elastomers because they act as an antioxidant (antirads) by scavenging the free radicals produced by irradiation. Hence quenching of free radicals initially produced during radiation will prevent further degradation caused by "oxygen mechanism" [31], [37] [72]. The percentage change in a property upon irradiation can be found by Equation 4.7.

$$\% \text{ change} = \frac{\text{Property after irradiation} - \text{Property before irradiation}}{\text{Property before irradiation}} * 100 \quad (4.7.)$$

Table 4.10 Mechanical properties of irradiated EPDM-CIIR nanocomposites

EPDM-CIIR Nanocomposites	Before Radiation	Low Dose 0.5MGy	Medium Dose 1MGy	High Dose 2MGy	Percentage change in TS		
					Low	Medium	High
					0.5MGy	1MGy	2MGy
E80CI20	1.39± 0.07	1.41±0.03	2.08±0.11	2.06±0.11	1.43	49.6	48.2
E80CI20NC2.5	1.85±0.09	1.92± 0.09	2.35±0.18	2.05±0.10	3.78	30.8	10.8
E80CI20NC5	2.21±0.10	2.42 ±0.12	2.69±0.13	2.39±0.12	7.52	21.7	8.14
E80CI20NC7.5	2.03±0.08	2.38± 0.12	2.59±0.13	2.24±0.11	17.2	27.6	10.3
E80CI20NC 10	1.97±0.08	1.98 ±0.09	2.27±0.11	0.72±0.04	0.51	15.2	-63.4

Amongst the EPDM-CIIR nanocomposites, E80CI20NC5 had the minimum percentage change in tensile strength and M100 with neat blend as compared to other filler levels. The percent

100

change in properties increased at higher filler loadings of nanoclay [237]. The nanocomposites tend to lose their elasticity when exposed to higher irradiation dosage of 2MGy due to chain scission which disintegrates the molecular chains [227], [228], [236]. Further the network contains more and more weakened zones after degradation, which deteriorates the material's ultimate properties.

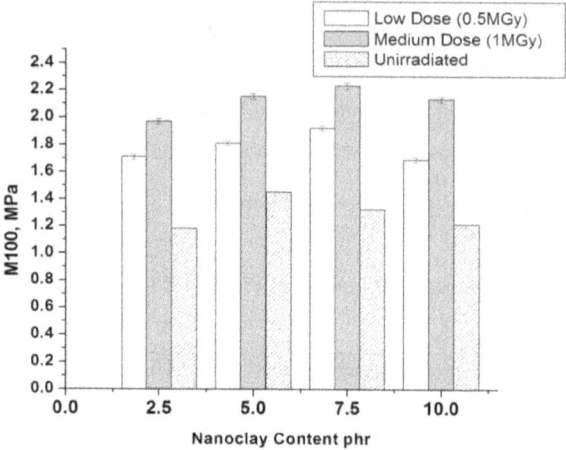

Figure 4.18 M100 of irradiated EPDM-CIIR nanocomposites

4.2.13 (c) Thermal degradation studies on irradiated nanocomposites

TGA was used to determine the thermal stability of irradiated nanocomposites. Initial degradation temperature (IDT) and peak degradation temperature of the irradiated nanocomposites were determined from TGA thermograms (Table 4.11). Increase in IDT at 0.5 MGy and 1 MGy indicate predominance of cross-linking while reduction in IDT at 2 MGy point towards chain scission [16], [227], [234]. These observations are consistent with those for obtained in FTIR analysis and mechanical properties. Similar trends were observed for all other nanocomposites.

Table 4.11 TGA properties of irradiated EPDM-CIIR nanocomposites

Radiation dose	Unirradiated	0.5 MGy	1 MGy	2 MGy
IDT (°C)	81	142	172	137
Peak degradation Temperature (°C)	461	472	472	461

4.2.13 (d) Transport coefficients after irradiation of nanocomposites

The sorption, diffusion and permeability coefficient decreased for 0.5 and 1 MGy due to the formation of cross-links. For higher dose of radiation, all the coefficients increased as the chain scission prevailed more. The sorption coefficients of irradiated E80CI20NC5 are tabulated in Table 4.12. Similar trends were observed for all other nanoclay contents. The well dispersed layered silicates in the EPDM-CIIR blend (at 5 phr) as evident from morphology, static and viscoelastic as well as solvent sorption behavior provided best gamma radiation ageing behavior compared to other nanoclay contents. The homogeneous distribution of nanoclay in E80CI20NC5 imparted tortous path as discussed from previous studies in section 4.2.11. This tortous path restricted the traverse of free radicals through the structure of nanocomposite. From Table 4.12, it can be concluded that nanoclay composites had better solvent barrier properties as compared to unfilled blends after exposure to gamma radiation [228], [230], [237]. This trend can be correlated with the mechanical properties and thermal analysis.

Table 4.12 Transport properties of irradiated nanocomposite (E80CI20NC5)

Radiation Dose	Un-irradiated	0.5 MGy	1 MGy	2 MGy
Diffusion coefficient $(Dx10^7)$ m^2/s	2.81	1.61	1.48	1.61
Sorption coefficient (S) (g/g)	1.35	1.32	1.29	1.35
Permeability coefficient $(P \times 10^7)$ (m^2/s)	3.79	2.12	1.90	2.17
Swelling coefficient (β) (cm^3/g)	1.73	1.69	1.65	1.74

The percentage change in swelling coefficient of irradiated nanocomposites and neat blends are tabulated in Table 4.13. The percentage change in swelling coefficient of E80CI20NC5 was minimal when compared to that of unfilled blend as evident from Table 4.13. Thus, the nanoclay based composites of EPDM-CIIR blends have potential application in rubber components exposed to radiation and hydrocarbon environments simultaneously.

Table 4.13 Comparison of percentage change in swelling coefficient after irradiation

EPDM-CIIR Nanocomposites	0.5MGy	1MGy	2MGy
E80CI20	33.8	35.8	34.9
E80CI20NC2.5	5.56	9.59	4.89
E80CI20NC5	2.30	4.60	0.05
E80CI20NC7.5	2.35	5.02	1.12
E80CI20NC10	33.2	34.7	34.1

4.3 Conclusions

This chapter provided detailed investigation of nanoclay based EPDM-CIIR nanocomposites for enhanced performance in radiation environments. The nanocomposites were prepared with varying organo-modified layered silicate contents (0, 2.5, 5, 7.5 and 10 phr) in a two-step solid state compounding process. From cure studies using oscillating disc rheometer, it was found that the scorch time and cure time of the nanocomposites increased with the increase in layered silicate content. FTIR studies of nanocomposites revealed chemical interaction between the nanoclay and the elastomer blend. Morphology studies (XRD and TEM) of the nanocomposites showed that the layered silicates were well dispersed in the elastomer matrix with intercalation and exfoliation up to 5 phr, though a few aggregates of layered silicates observed at higher filler level (10 phr). The static and viscoelastic behavior of nanocomposites of EPDM blends were investigated with varying organo modified layered silicates (OMLS) or nanoclay content.. The nano-reinforcement of EPDM-CIIR blends enhanced static mechanical properties till 5 phr owing to the larger interfacial interactions between the nanoclay and rubber as well as uniform dispersion in the blend and decreased thereafter due to formation of aggregates. The slight decline in properties for nanocomposites at 10 phr nanoclay can be attributed to the reduction in interfacial area due to agglomeration. Mooney-Rivlin models also established the non-linear reinforcement effect of nanofiller. Rheometric characteristics were used to evaluate optimum cure time and Mooney viscosities. Incorporation of OMLS increased the stiffness of the blend. Exploration of dynamic mechanical properties showed noteworthy enhancement in storage modulus in the nanocomposites below the glass transition temperature. The reinforcement factor, degree of entanglement and constrained volume regions were evaluated from dynamic mechanical analysis. The experimentally obtained static modulus was compared with those predicted by several micro-mechanical models. The relative storage modulus predicted by Liang model was matching with experimental results in blends exhibiting exfoliated morphology. The viscoelastic behavior of nanocomposites were studied from filler-rubber interactions under dynamic strain sweep conditions and stress relaxation tests, to understand the performance of elastomer in continuous load bearing applications like seals. The shear storage modulus increased and loss tangent under strain sweep conditions decreased with increase in nanofiller content. The applicability of Kraus model to storage modulus obtained from Payne effect was studied. The elasticity and stress relaxation for the nanocomposites were analyzed and found that the maximum delay was for blends with well dispersed silicates. The storage modulus obtained from Payne effect were deviated from Kraus model prediction because of smaller size of aggregates in nanocomposites when compared to microfillers. The solvent sorption behavior of nanocomposites in cyclohexane was determined and found that the solvent uptake declined with the increase in filler content. The transport coefficients such as sorption, diffusion, and permeation were evaluated and found to decrease with increasing nanoclay content owing to the rise in tortuosity, increase in cross-link density and reduction in free volume. From TGA studies, it was evident that EPDM-CIIR nanocomposites had higher peak degradation temperature indicating better thermal stability after reinforcement. The influence of gamma irradiation on EPDM/CIIR nanocomposites was evaluated after exposing to ^{60}Co source gamma rays for cumulative doses upto 2 MGy. Up to 1 MGy of radiation dosage, cross-linking is predominant whereas at 2 MGy chain scission was predominant in the nanocomposites. The tortuous path for migration of free radicals created by uniform intercalation and exfoliation of layered silicate in the elastomer and free radical scavenging ability of the nanofiller minimized the damage from γ radiation. The least change in mechanical properties and sorption coefficients were observed for blend with 5 phr nanoclay. FTIR and ESR spectroscopy gave an insight to chemical changes and presence of free radicals respectively after irradiation. The thermal degradation behavior after irradiation was similar to mechanical and transport properties as well as spectroscopic analysis.

Chapter 5

Silane modified nanosilica particles reinforcement on EPDM-CIIR blends for gamma radiation environment[4]

5.1 Introduction

The influence of the influence of bis(3-triethoxysilylpropyl)tetrasulfide (TESPT) grafted nanosilica (NS) reinforcement on the rheometric characteristics, mechanical and viscoelastic behavior, solvent barrier properties and thermogravimetric studies as well as behavior after exposure to different cumulative γ-radiation doses of EPDM-CIIR blends is studied in this chapter. Amongst the nanoparticles, nanosized silicon dioxide or nanosilica (SiO_2), is a good candidate for reinforcement of rubbers owing to their high stability, lower toxicity and capability to be modified with a range of functional groups. The nanosilica particles are chemically modified with TESPT and the formation of covalent linkages upon surface modification of nanosilica was observed from FTIR spectroscopic analysis. The physico-chemical interactions of modified silica nanoparticles ($mSiO_2$) and rubber chains have been characterized with morphology and spectroscopic studies. The suitability of analytical models for sorption behavior has been evaluated in this chapter to investigate the type of diffusion. The kinetic aspect of sorption behavior is also explained in this chapter. The activation energy (E_a) for thermal degradation was calculated by Horowitz-Metzger and Coats-Redfern methods from TGA data. The effect of three different cumulative doses of gamma radiation on nanosilica reinforced EPDM-CIIR blends are also explained in this chapter.

---

[4] This chapter has been published in *Journal of Applied Polymer Science* (Neelesh Ashok *et al.*, "Synergistic enhancement of mechanical, viscoelastic, transport, thermal and radiation ageing characteristics through chemically bonded interface in nanosilica reinforced EPDM-CIIR blends", *2020,* DOI: 10.1002/app.50082)

5.2 Results and discussions

5.2.1 Functionalization of nanosilica and mSiO$_2$-rubber interactions

The functionalization of nanostructured silica particles can introduce various functional groups which can enhance hydrophilicity to attach organic polymer chains by establishing covalent bonds. Seyed *et al.* [238] investigated on various types of silane modifier on the surface of nanosilica particles and role of silica-rubber interfacial interactions on vulcanization, morphology, mechanical properties and viscoelastic characteristics of rubber composites.

Interfacial interactions generated as a result of formation of rigid covalent bonds from bifunctional silanes significantly improved mechanical behavior and filler networks in the rubber matrix [114]. The hydroxyl (–OH) groups on the surface of nanoSiO$_2$ particles form hydrogen bonds and tend to form agglomerates. Hence when nanoSiO$_2$ particles are directly fed into the polymer, the filler-filler network formed due to hydrogen bonds makes it difficult to attain uniform dispersion in the matrix.

The formation of hydroxyl groups are due to strong attraction between silanol groups in the surface of nanoSiO$_2$ particles. The silica tactoids are hard to disagglomerate and they also tend to flocculate in the filled compounds after secession of shear during curing process [239]. Surface modification of nanoSiO$_2$ particles can reduce formation of such aggregates [239].

TESPT contains organic functional group and hydrolysable silyl group. The polar group of the TESPT (triethoxysilylpropyl), silyl group reacts with oxygen on the surface of nanoSiO$_2$ by electrostatic interactions during surface modification. The FTIR spectra and schematic representation of nanoSiO$_2$ before and after treatment and pure TESPT are shown in Figure 5.1.

It was evident that two new peaks found at 2178 cm^{-1} and 2448 cm^{-1} corresponds to formation of Si-H and Si-CN$^-$ bonds. The grafting mechanism in SiO$_2$-TESPT can be attributed to appearance and increase in intensity of peak at wavenumber 2734 cm^{-1}- 2930 cm^{-1} corresponding to–CH bond stretching in methylene group which is present in TESPT. An increase in depth of peaks at 1633cm^{-1} and 1881 cm^{-1} depicted rise in carbonyl group (C=O) which is a characteristic of silane coupling agent. The coupling agent chemically interfaces with silanol group in nanoSiO$_2$ and rubber chains.

The chemical reaction occurring between EPDM/CIIR elastomer chains, TESPT and nanoSiO$_2$ are shown in Figure 5.2. The sulfur present in TESPT interacts with the polymer chains as depicted in Figure 5.2.

After surface treatment the silica particles get dispersed easily into the matrix and have better anchoring sites with polymer chains. The formation of new bonds for silane functionalized nanosilica particles established better chemical interaction with EPDM-Chlorobutyl rubber chains.

The nanosilica was modified using silane coupling agent (TESPT). The Si-69 or silane coupling agent is primarily responsible for enahced adhesion as discussed above by initiating coupling reaction between inorganic filler and organic elastomer. The nanosilica powder are mixed with TESPT for 30 minutes and kept in oven at 110°C followed by drying at room temperature for 24 hours.

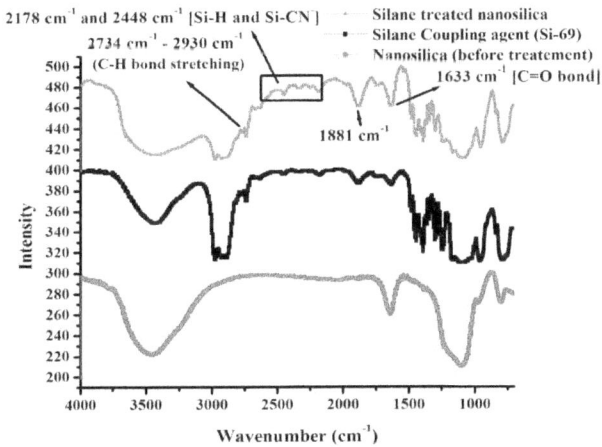

Figure 5.1 FTIR spectra of nanosilica (before and after modification) and Si-69

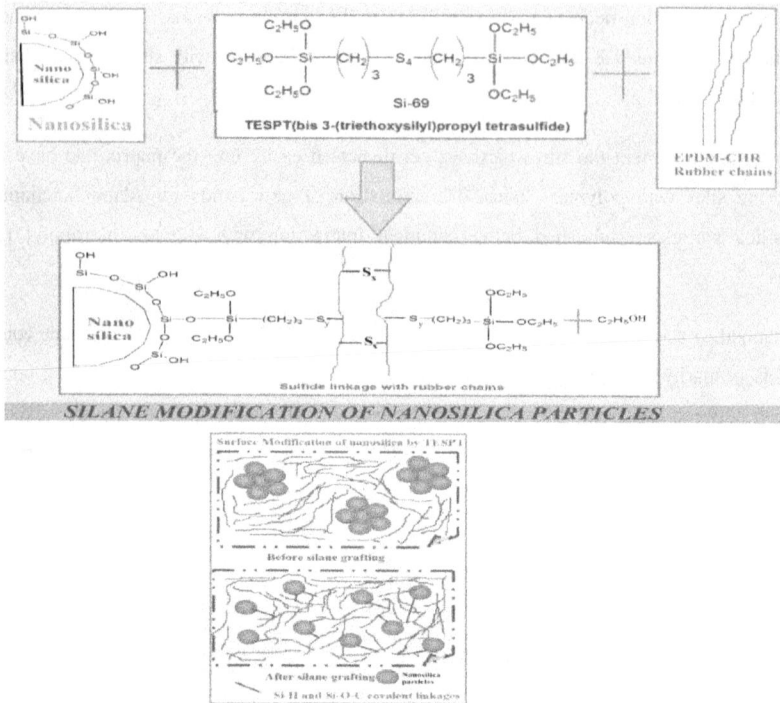

Figure 5.2 Chemical reaction and schematic representation after silane modification

5.2.2 Rheometric Characteristics

The rheometric characteristics of $mSiO_2$ reinforced blends and E80CI20 are given in Table 5.1. The increase in τ_{min} for all nanocomposites substantiates the enhancement in viscosity with increasing the $nanoSiO_2$ loading. This enhancement was attributed to formation of physical crosslink networks between filler and rubber. Improvement in τ_{min} also indicated that large numbers of rubber chains are immobilized near filler-rubber interface regions thus establishing reinforcement effect [240].

The τ_{max} can be considered as the highest torque obtained from ODR. The torque is equivalent to stiffness of completely cross linked rubber and is proportional to crosslink density. As summarized in Table 5.1, τ_{min} and τ_{max} increased as a function of nanosilica loading. t_{90} is defined as time required for attaining 90% of maximum torque.

The decrease in t_{90} and increase in CRI for $mSiO_2$ filled blends was because of the presence of functional groups like siloxane and acidic hydroxyl on surface of $mSiO_2$ particles [238]. These groups produced after surface modification of silica were able to accelerate the degree and rate of curing by providing active sites for curing [241].

Apart from these groups, the elemental sulfur present in TESPT structure also has the ability to participate in cure reaction thus reducing t_{90}. The undesirable effect of silanol groups in cure reaction was eliminated after surface modification.

Table 5.1 Rheometric characteristics of nanosilica reinforced EPDM-CIIR blends

Rheometric characteristics	E80CI20	E80CI20NS2.5	E80CI20NS5	E80CI20NS7.5	E80CI20NS10
Optimum cure time t_{90} (min)	7.55	6.90	6.07	5.88	6.33
CRI	16.0	22.7	23.1	19.7	17.5
Minimum torque τ_{min} (Nm)	1.67	2.45	2.51	2.76	2.64
Maximum torque τ_{max} (Nm)	3.23	4.67	5.15	5.54	4.19
(τ_{max}-τ_{min}) (Nm)	1.56	2.22	2.64	2.78	1.55
Initial Mooney viscosity (IV)	56.5	73.2	73.4	74.4	93.4
Mooney Viscosity (FV)	25.6	37.3	33.9	31.6	32.7
Delta Mooney (IV-FV) (Mooney)	30.9	35.9	39.5	42.8	60.7

Mooney viscosity provides an insight into the cure rate of an elastomer from the measured torque. As evident from Table 5.1, Mooney viscosity increased as a function of nanoSiO$_2$ content [242].

Generally for sulfur cured elastomers, the addition of mSiO$_2$ increases the torque due to more filler-rubber interactions obtained from alkoxy-silanol and rubber-sulfide linkages. More torque is required for the rubber chains to slip past each other at the constrained regions near filler interface. The increase in delta Mooney (difference between initial value (IV) and final value (FV) of torque) upon incorporation of filler also provided evidence on rise in stiffness of nanocomposite and reinforcing effect of mSiO$_2$.

5.2.3 FTIR Analysis

The FTIR spectra of E80CI20 and mSiO$_2$ reinforced blends are shown in Figure 5.3. The depth of peaks at 1031 cm^{-1}, 1369 cm^{-1}, 1456 cm^{-1} and 1538 cm^{-1}corresponding to Si-O-Si (Stretching and flexural vibrations), C-H bending (–CH$_2$ and –CH$_3$ groups) and C-C bonds increased as a function of mSiO$_2$ content. The increase in siloxane linkages reflected from FTIR spectra provided confirmation on occurrence of chemical interactions between the surface of mSiO$_2$ and the blend [189].

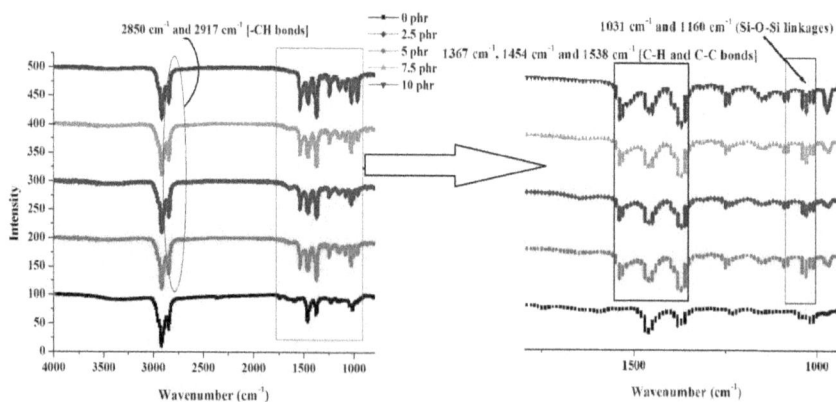

Figure 5.3 FTIR spectra of nanosilica reinforced EPDM-CIIR blends

111

5.2.4 TEM analysis

The reinforcement efficiency in rubbers is based on the extent of dispersion and distribution of nanoparticles in the blend matrix. As an illustration, TEM micrographs of 5, 7.5 and 10 phr mSiO$_2$ incorporated blends are shown in Figures 5.4 (a), (b-c) and (d) respectively. Homogenous distribution of SiO$_2$ nanoparticles in EPDM-Chlorobutyl rubber matrix was seen in TEM micrograph of E80CI20NS7.5 and E80CI20NS5. In TEM micrograph of E80CI20NS10, few aggregates of SiO$_2$ particles were also seen along with disseminated mSiO$_2$ particles. A good distribution of SiO$_2$ nanoparticles in EPDM-CIIR matrix after surface modification was also evident from TEM micrographs.

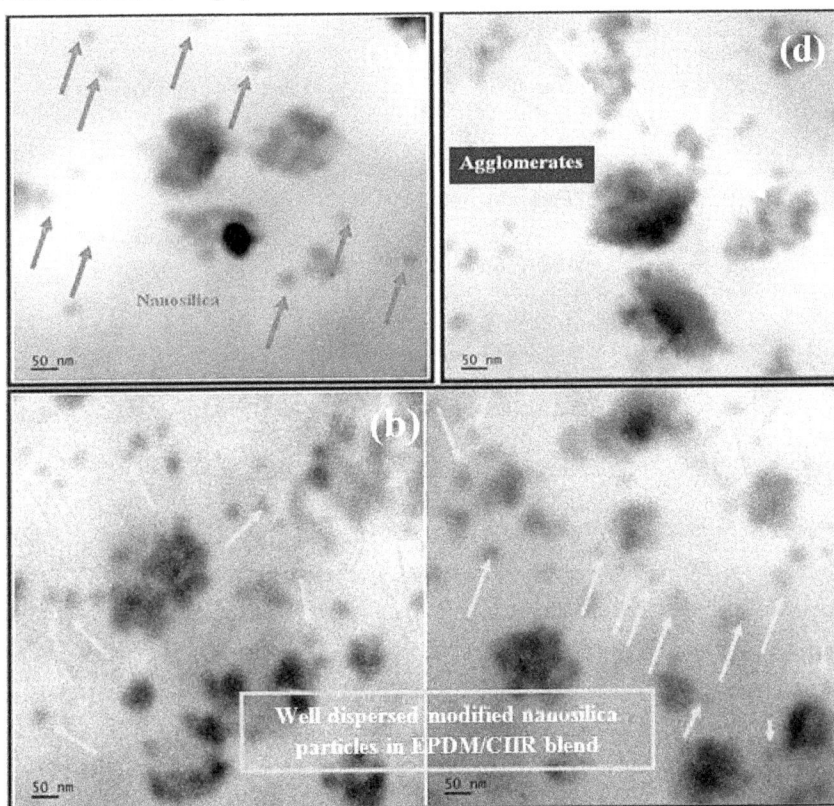

Figure 5.4 TEM micrographs of (a) E80CI20NS5 (b-c) E80CI20NS7.5 and (c) E80CI20NS10

5.2.3 Mechanical properties

The reinforcement of EPDM-Chlorobutyl rubber blends with $mSiO_2$ gave rise to a noticeable improvement in mechanical properties which are tabulated in Table 5.2. The tensile strength and modulus at 100 percent elongation (M100) increased upto 7.5 phr $mSiO_2$. A slight decline in properties was observed for EP80CI20NS10 due to formation of spherical agglomerates observed from TEM micrographs. The incorporation of 7.5 phr $mSiO_2$ in the matrix showed maximum rise in tensile strength and elastic modulus, by 63.7% and 118% respectively.

Table 5.2 Static tensile properties of nanosilica reinforced EPDM-CIIR blends

Sample	Tensile strength MPa	% Elongation at break E_b	M100 MPa	Percent increase in TS	Percent increase in M100	$\Delta G = (G_o' - G_\infty'),(1-300\%$ strain range)
E80CI20	1.38 ± 0.07	97 ± 6	0.92 ± 0.05			221
E80CI20NS2.5	1.72 ± 0.05	114 ± 2	1.57 ± 0.04	24.6	70.6	325
E80CI20NS5	1.98 ± 0.08	122 ± 3	1.86 ± 0.02	43.4	102	367
E80CI20NS7.5	2.26 ± 0.11	135 ± 4	2.01 ± 0.03	63.7	118	419
E80CI20NS10	1.59 ± 0.10	109 ± 3	1.55 ± 0.02	15.2	68.4	377

The evenly distributed SiO_2 nanoparticles establish larger interface with the elastomer chains. The entanglement of elastomer chains on the filler and strong interfacial $mSiO_2$-rubber interactions restricted the mobility of chains. The larger interface also facilitates effective transfer of applied stress to $mSiO_2$ particles. These aspects contribute to enhanced reinforcement and mechanical properties in $mSiO_2$ reinforced EPDM-Chlorobutyl blends. The stress-strain response of nanosilica based nanocomposites of EPDM blends are plotted in Figure 5.5. Well distributed $mSiO_2$ in the EPDM- Chlorobutyl rubber matrix reduces inter-particle distance, resulting in overlap of immobile elastomer chains at the interface regions. The addition of nanoparticle can generate three regions (interfacial layers or zones) as depicted in Figure 5.6.

The region adhering to the nanoparticle is the rigid nano layer where the segmental mobility of polymer chains is restricted. In the next region, polymer chains are bounded tightly to the rigid phase and are called as constrained zone/rigid-amorphous zone [114], [243]. The outer region is the free zone where the polymers chains are free to slip past each other. The silica-rubber interface substantially improved by effective silanization as discussed in the previous section, forms strong nano layer of immobilized EPDM-CIIR chains around the silica surface. The reduction in mechanical properties of EP80CI20NS10 was due to formation of $mSiO_2$ aggregates as evident from TEM shown in Figure 5.4(d). The agglomerates of $mSiO_2$ particles cause non-uniform distribution of fillers and reduction in silica-rubber interfacial interactions.In addition, the agglomerates of silica particles develop more stress concentration at the filler-rubber interfaces limiting the efficiency in load transfer from matrix to filler. The improved static mechanical behavior in nanoSiO2 reinforced EPDM-Chlorobutyl rubber blends also provide an indirect measure of good physico-chemical interactions viz. vander waals forces, chemical bonds etc. between surface of silica particles and matrix. Yoshinobu *et al.* [244], [245]reported that reinforcement effect of $mSiO_2$ particles in rubber was influenced strongly by entanglement of chains on the modified silica surfaces and crosslinking density. Ishida and coworkers [97], [195], [204] named "interfacial layer in the interphase" which acts as interphase locking sites, consists of both silane modifiers chemically bonded on filler surface and elastomer chains [246], [247].

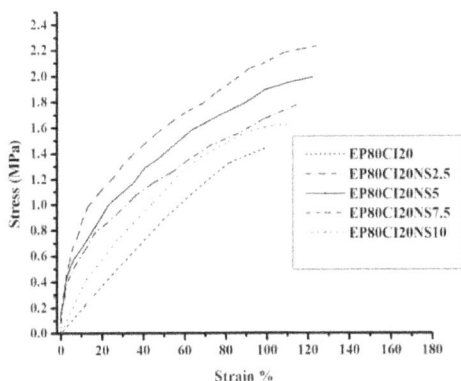

Figure 5.5 Stress-strain response of nanosilica reinforced EPDM-CIIR blends

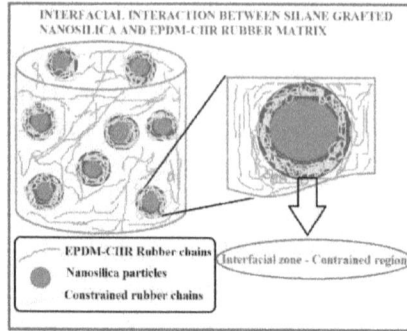

Figure 5.6 Schematic representation of interfacial interactions of nanosilica reinforced blends

5.2.4 Payne effect (filler-rubber interactions) and stress relaxation (force decay)

The viscoelastic behavior of $mSiO_2$ reinforced EPDM-Chlorobutyl rubber blends are dependent on the interactions occurring in elastomer nanocomposites viz. filler-filler and rubber-filler interactions [248]. The results obtained from strain-sweep measurements demonstrate the features attributing to Payne effect [249]. The effect of strain sweep on storage moduli(G') for different $mSiO_2$ loading is shown in Figure 5.7(a).The storage modulus was at peak value (referred as G_o') at lower strains and gradually decreased at higher strains (referred as G_∞'). The highest G_o' observed at lower strains were due to filler-rubber interactive forces produced from nanoreinforcements [250]. The amplitude of Payne effect ($\Delta G = G_o' - G_\infty'$) is dependent on the structural and surface characteristics, specific surface area and distribution of filler in the matrix. Here, G_o' is taken at 1% dynamic strain and G_∞' at 300% strain.

Increase in ΔG with increase in $mSiO_2$ content, as tabulated in Table 5.2, proved that Payne effect was significant in the nanocomposites. The G' increased till 7.5 phr of $mSiO_2$content as observed from Figure 5.7(a). This behavior was similar to trend observed in static mechanical and cure characteristics. The larger $nanoSiO_2$-rubber interfacial interaction observed in E80CI20NS7.5 provides more anchoring of rubber chains at filler interface as shown in Figure 5.7(a).When comparing the degree of Payne effect amongst the nanocomposites, the blend with 10 phr SiO_2 content has lower value than 7.5 and 5 phr SiO_2 due to formation of agglomerates as observed from TEM analysis. The breakdown of filler-filler aggregates occurs at higher strains.

115

Stress relaxation or structural viscosity tests were carried out to evaluate the deformation of elastomeric materials with respect to time [225]. In stress relaxation analysis, the variation in the decline of stress with time as a function of $mSiO_2$ particles was analyzed. The decline in stress/modulus with time for $mSiO_2$ reinforced EPDM- Chlorobutyl rubber blends are shown in Figure 5.7(b). The modulus decay was due to the physical changes incurred when the material is strained at initial and final zones. The time-dependent physical relaxation in rubber takes place due to rearrangement of molecular chains and fillers. It is important to note that the relaxation occurs in two phases because of different types of interactions present in elastomer nanocomposites [251].

The initial phase relates to breakage of filler-filler networks and final phase indicated a rupture of jammed filler-polymer networks as represented in Figure 5.7(b). It can be observed that blend with well dispersed $mSiO_2$ (7.5 phr) has lower stress relaxation rate. Well distributed fillers in the matrix impart more reinforcement efficiency thus lowering the rate of stress relaxation. The breakdown of weak filler-filler aggregates in E80CI20NS10 created large rate of stress relaxation. The least rate of stress relaxation seen for E80CI20NS7.5 provides better structural integrity over time which is better for continuous load bearing applications.

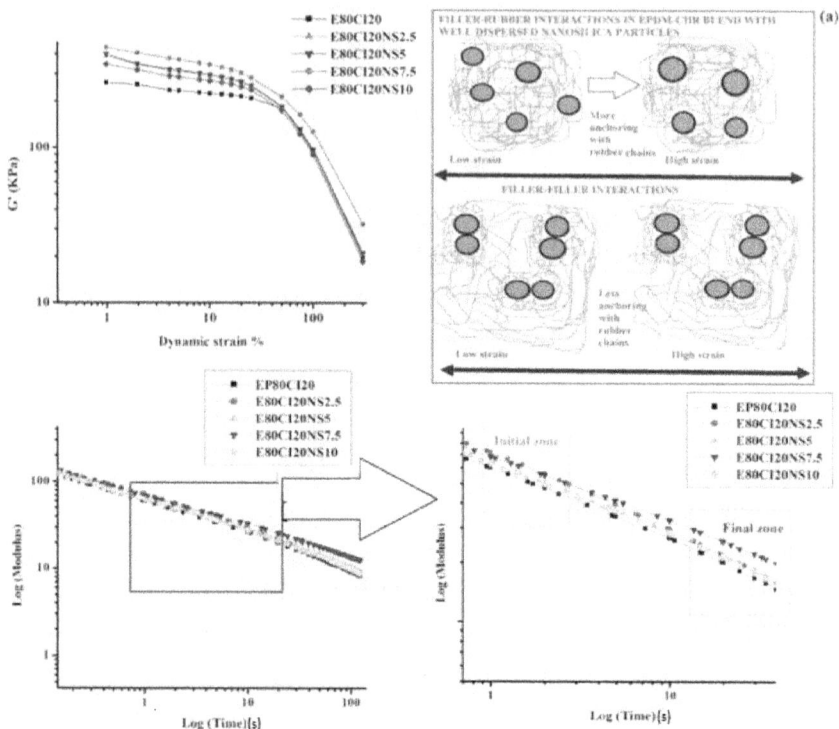

Figure 5.7 (a) Payne effect and (b) Stress-relaxation of nanosilica reinforced EPDM-CIIR blends

5.2.5 Solvent permeation characteristics

The solvent permeation characteristics of $mSiO_2$ reinforced EPDM-Chlorobutyl rubber blends are analyzed to understand the performance of nanocomposites in resisting hydrocarbon solvents. The swelling of polymer is a sorption process which involves the permeation of solvent from one part to another [87]. The solvent permeation is a kinetic process which is dependent on several parameters like free volume in the matrix, dynamics or segmental mobility of polymer chains and size of penetrate molecule. The solvent permeation takes place in three steps initially molecules are absorbed in the surface of rubber, then molecules traverse through the rubber chains and finally the solvent molecules desorb at the surface of the rubber nanocomposite. The deviation in solvent uptake upon increase in $mSiO_2$ content in the blends was evaluated using cyclohexane as the solvent. As seen in previous chapter, if nanosized fillers are incorporated in the matrix, it will take up the free volume within the elastomer blend and imparts a 'tortous' path for solvent permeation. The transport behavior can be used to provide an insight into the morphology of nanocomposites [68], [101].

The sorption curves were plotted with the number of moles of solvent uptaken per 100g of rubber (Q_t) vs $\sqrt{t}$ for EPDM- Chlorobutyl rubber blend reinforced with nanosilica as shown in Figure 5.8. It was revealed that the absorption of solvent reduced for nanoreinforced samples than pristine rubber blend. Due to the availability of maximum free volume and increased segmental mobility in neat rubber, the solvent uptake was highest in E80CI20 as compared to nanofilled blend. The solvent absorption was least for blends with well distributed and dispersed nanosilica particles (E80CI20NS7.5). It is due to the fact that dispersed nanosilica particles embedded in rubber blend imparting solvent permeation resistance. Another reason for less solvent uptake is that mobility of rubber chains gets restricted due to strong filler-polymer interactions as explained in the previous sections. The rubber chains become less flexible leading to low sorption behavior. Thus, by incorporating $mSiO_2$ in the EPDM/CIIR matrix, the solvent permeation through the nanocomposite gets reduced.

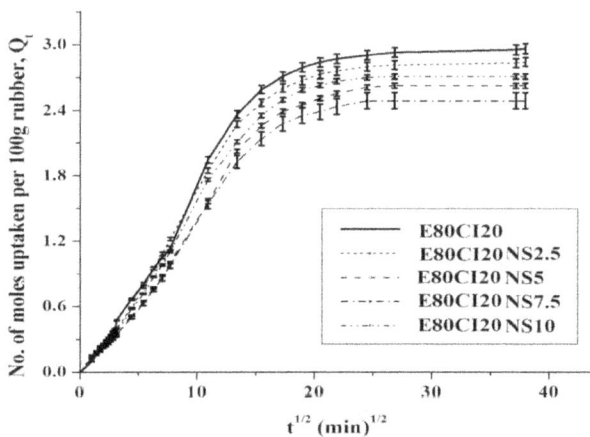

Figure 5.8 Sorption isotherms of nanosilica reinforced EPDM-CIIR blends

118

In the sorption curve, two distinct zones are observed corresponding to initial solvent uptake and reduced swelling rate tending towards equilibrium as discussed in previous chapter. In the initial zone, the rate of solvent sorption is very high due to large concentration gradient and the rubber is under solvent stress. The swelling rate gradually decreases in the secondary zone due to decrease in concentration gradient tending towards equilibrium. At higher nanofiller content (10 phr) the solvent uptake is more due to the aggregation of filler. As a result of agglomeration at some portions, nanoparticles are not available in the rubber matrix to enhance the extended path length for solvent migration [114]. The transport coefficients (Permeation, Diffusion and Sorption) of nanocomposites calculated from sorption curve are tabulated in Table 5.3.

The value of diffusion coefficient decreased with increase in $mSiO_2$ content as documented in Table 5.3. Apart from reduction in free volume in matrix upon reinforcement as stated earlier, the reason for reduction in D is due to restriction in rubber chain mobility at nano interfaces as explained in previous sections. The sorption coefficient (S) value is the maximum saturation sorption in equilibrium and is calculated by the following equation

$$\text{Sorption coefficient (S)} = \frac{\text{Swollen weight in equilibrium}}{\text{Initial weight}} \quad (5.1.)$$

The intensity of interactions between the solvent and matrix is known from sorption coefficient. It is estimated from plateau region in the sorption curve of EPDM-Chlorobutyl rubber nanocomposites and is presented in Table 5.3. The decrease in S upon nanoreinforcement also confirms the reduction in solvent uptake upon nanoreinforcement.

The permeation coefficient (P) point towards the quantity of solvent permeated through uniform area of sample per second. The following equation gives the calculation of permeation coefficient

$$\text{Permeation coefficient (P)} = \text{Diffusion coefficient (D)}* \text{Sorption coefficient (S)} \quad (5.2.)$$

For all the nanocomposites analyzed, the permeation coefficient decreased as a function of $mSiO_2$ content. The maximum values for diffusivity and permeability were obtained for neat blends and minimum for the nanocomposite with well exfoliated filler (7.5 phr). The long-range mobility of elastomeric chains gets restricted after the vulcanization process, but the local segmental mobility remains unaffected. When the reinforcing filler like modified nanosilica particles are added on to the matrix, the local mobility of EPDM-Chlorobutyl rubber chains gets restricted and thus solvent permeation barrier gets enhanced.

119

Table 5.3. Transport properties and modeling of nanosilica reinforced EPDM-CIIR blends

Sample	Diffusion Coefficient (D x 10^7) m^2s^{-1}	Sorption Coefficient (S) (gg^{-1})	Permeability Coefficient (P x 10^7) m^2s^{-1}	Swelling Coefficient (β) (cm^3g^{-1})	Crosslink Density x 10^4 (gmol cm^{-3})
E80CI20	3.51	1.59	5.58	2.04	4.05
E80CI20NS2.5	3.45	1.49	5.14	1.92	4.42
E80CI20NS5	2.93	1.43	4.18	1.84	4.98
E80CI20NS7.5	2.75	1.31	3.60	1.69	5.24
E80CI20NS10	3.39	1.45	4.91	1.91	4.61

Analytical Models	Korsmeyer-Peppas model		Peppas - Sahlin model				Higuchi model	
	R^2/χ^2	n	R^2/χ^2	K_f	K_r	m	R^2	K_h
E80CI20	0.99/0.13	1.19	0.98/0.06	0.01	0.35	0.16	0.98	0.05
E80CI20NS2.5	0.99/0.11	1.10	0.99/0.05	0.01	0.30	0.17	0.98	0.05
E80CI20NS5	0.99/0.10	1.09	0.99/0.06	0.10	0 16	0.20	0.98	0.05
E80CI20NS7.5	0.99/0.14	1.08	0.99/0.05	0.13	0.17	0.18	0.98	0.05
E80CI20NS10	0.99/0.12	1.05	0.98/0.08	0.08	0.31	0.15	0.98	0.05

5.2.5 (a) Sorption kinetics

The kinetic aspect of sorption behavior was analyzed by fitting the solvent sorption data to the following equation [176]

$$\ln(\frac{Q_t}{Q_\infty}) = \ln(k) + n*\ln(t) \qquad (5.3)$$

where Q_t is the percentage number of moles of solvent uptaken at time t, Q_∞ is the solvent uptake at equilibrium and k is a constant which is related to the structural characteristics of EPDM-Chlorobutyl rubber matrix and its interaction with the solvent. The value of n, provides an idea about mechanism behind solvent sorption behavior viz. Fickian, Non-Fickian and Anomalous [89]. The values of n can be 0.5, 1 and in between 0.5 and 1 based on the type of former mentioned mechanism. The values of n and k for $mSiO_2$ reinforced EPDM/CIIR blends tabulated in table were determined by linear regression analysis. From Table 5.4, it can be observed that the value of n is close to 0.5 indicating the sorption mechanism is converging Fickian. This can be attributed to the fact that rate of diffusion of solvent molecule is lower than that of elastomer chain relaxation in $mSiO_2$/EPDM/CIIR composites.

In addition, the value of k is found to be higher for $mSiO_2$ filled blends compared to neat ones. The higher value of k for nanocomposites also proves the increase in filler-rubber interactions. The effect of temperature on solvent permeation was also studied for $mSiO_2$ reinforced EPDM-Chlorobutyl rubber blends. The transport properties were evaluated at 30°C and 45°C using cyclohexane as solvent. The increase in solvent uptake with rise in temperature can be ascribed to enhanced segmental mobility of elastomeric matrix and improved kinetic energy of the solvent molecules resulted from augmented number of collisions at high temperature. The progress in flexibility of polymer chains at higher temperature also facilitates solvent permeability. Hence it can be inferred that phenomenon of solvent permeation is temperature dependent.

The Arrhenius equation is used for calculating energy required for the diffusion or permeation of solvent molecule [100],

$$D=D_0 e^{-E_D/RT} \quad (5.4.)$$
$$P=P_0 e^{-E_P/RT} \quad (5.5.)$$

where Do and Po are diffusion and permeation coefficients (extrapolated to zero permeation concentration) respectively. T is the temperature in Kelvin and R is the gas constant. The energies of activation for diffusion and permeation are represented by E_D and E_P respectively. The activation energy for diffusion (E_D) was calculated from the slope of ln D versus 1/T plot. The activation energy needed for diffusion increased as a function of $mSiO_2$. The activation energy for permeation (E_P), also showed similar behaviour as E_D. The Van Hoff equation was used to calculate enthalpy of sorption (ΔH_S)

$$E_P=\Delta H_S+E_D \quad (5.6.)$$

The value of enthalpy increased with $mSiO_2$ content indicating that the sorption characteristics were an exothermic process.

121

Table 5.4 Sorption kinetics of nanosilica reinforced EPDM-CIIR blends

Sample	30°C				45°C			
	Swelling coefficient (β) (cm³g⁻¹)	Sorption coefficient (S) (g g⁻¹)	Diffusion coefficient (D×10⁷) (m² s⁻¹)	Crosslink Density υ × 10⁴ (gmol cm⁻³)	Swelling coefficient (β) (cm³g⁻¹)	Sorption coefficient (S) (g g⁻¹)	Diffusion coefficient (D×10⁷) m²/s	Crosslink Density υ x 10⁴ (gmol cm⁻³)
E80CI20	3.01	3.35	4.54	1.71	2.99	3.31	5.25	1.75
E80CI20NS2.5	2.94	3.29	4.32	1.69	2.93	3.28	5.19	1.64
E80CI20NS5	2.72	3.11	3.93	1.61	2.81	3.19	4.72	1.58
E80CI20NS7.5	2.54	2.98	3.67	1.49	2.56	2.99	4.61	1.48
E80CI20NS10	2.61	3.27	4.72	1.43	2.48	2.93	6.14	1.75

Kinetic parameters	30°C		45°C		Energies of permeation and diffusion		
	n	k	n	k	E_D (kJ mol⁻¹)	E_P (kJ mol⁻¹)	ΔH_S (kJ mol⁻¹)
E80CI20	0.51	0.04	0.50	0.069	8.99	9.55	-1.23
E80CI20NS2.5	0.52	0.04	0.50	0.071	9.78	9.62	-1.59
E80CI20NS5	0.54	0.04	0.52	0.054	10.2	11.9	-1.73
E80CI20NS7.5	0.53	0.04	0.53	0.061	11.7	12.1	-3.67
E80CI20NS10	0.53	0.04	0.49	0.067	10.3	7.93	-2.37

5.2.5 (b) Theoretical modeling of permeation characteristics

The permeation characteristics of nanocomposites are modeled with respect to composite theories of solvent permeation and experimental data are compared with analytical values. The theoretical models also aid to evaluate the mode of transport through the rubber from diffusion kinetics data. The diffusion parameters of the nanosilica reinforced EPDM blends were compared with Korsmeyer-Peppas model, Peppas-Sahlin equation and Higuchi model. These models are dependent on process in which solvent molecules migrate from their initial position to peripheral surface of matrix.

The solvent permeation path is affected by the physico-chemical characteristics of the penetrant and the structure of the elastomer system. The solvent uptake values of nanocomposites obtained experimentally were fitted with above mentioned models. Solvent diffusion and matrix swelling are the two major forces governing transportation phenomena [90], [117], [223].

Korsmeyer-Peppas model

The Korsmeyer-Peppas model is expressed as

$$\frac{Mt}{M\infty} = k\,t^n \quad (5.7)$$

Where k is the constant distinctive of filler-polymer system, n represents the diffusion constant and $\frac{Mt}{M\infty}$ is the fraction of solvent released at time t. The diffusion mechanism is dependent on values of n – Fickian diffusion mechanism is specified when n<0.45, non-Fickian diffusion or anomalous transport is indicated when n lies between 0.45 and 0.89 and case II and super case II transport mechanism is pointed out when n is 0.89 or above.

Peppas-Sahlin model

According to Peppas-Sahlin equation, diffusion mechanism in polymers is a combination of two process ie. diffusion in the swollen polymer matrix and relaxation of polymer chains in the matrix. The Peppas-Sahlin equation is as follows

$$\frac{Mt}{M\infty} = K_f t^m + K_r t^{2m} \quad (5.8)$$

Where K_r is the relaxation coefficient, K_f is the diffusion fickian coefficient, $\frac{Mt}{M\infty}$ represents the fraction of solvent released at time t and m is the Fickian diffusion exponent. When K_f and K_r are equal, then the diffusion mechanism is the combination of both diffusion and polymer chain relaxation. The solvent release is mainly controlled by diffusion when K_f is greater than K_r and when the release is result of matrix swelling K_r is more than K_f.

Higuchi model

The Higuchi model is formulated to evaluate the rate of solvent release based on laws of diffusion. This first order model for understanding diffusion in polymers is given by the following relation

$$\frac{Mt}{M\infty} = k\, t^{0.5} \qquad (5.9)$$

Where t is time, M_t is molar solvent uptake and k is the Higuchi constant. This model is based on the following assumptions (1) unidirectional state of diffusion (2) penetrate particles are much smaller than matrix system (3) diffusivity is a constant and (4) dissolution and matrix swelling are constant. Theoretical model fitting values of sorption correlation coefficients are tabulated in Table 5.3. Among the models, Korsmeyer-Peppas model have shown the values of n>1 for all the samples, this indicated super case II transport phenomenon. From Peppas-Sahlin model it was confirmed that the diffusion was due to matrix swelling because K_r values were observed higher than that of K_f from Table 5.3.

5.2.6 Thermal degradation properties

The thermal stability and degree of thermal degradation of $mSiO_2$ reinforced EPDM-Chlorobutyl rubber blends were determined from thermo gravimetric analysis. The thermal properties of nanocomposites are tabulated in Table 5.5. The temperatures at initial stage of degradation (T_i), degradation at peak rate (T_f) and terminal degradation (T_o) increased for $mSiO_2$ reinforced blends when compared to unfilled blends. Thermal decomposition profiles (TGA and DTG curves) of $mSiO_2$ reinforced and neat EPDM-CIIR blends are shown in Figure 5.9. Amongst the nanocomposites, E80CI20NS7.5 has highest shift towards right side in comparison with E80CI20. The well distributed nanoSiO2 particles in the matrix improved thermal stability. The shift in thermal degradation to higher temperature was observed for $mSiO_2$ reinforced blends because the elastomer chains constrained at the $mSiO_2$ interface degrade slowly. The presence of thermally stable $mSiO_2$ particles improved the thermal stability of the nanocomposite. In addition, the $mSiO_2$ particles promote uniform dissipation of heat in the structure of nanocomposite which could also enhance thermal properties. The formation of nano agglomerates in E80CI20NS10 lowered thermal stability. Many unconstrained elastomer chains in E80CI20NS10 tend to degrade faster reducing thermal stability.

Table 5.5 Thermal degradation properties of nanosilica reinforced EPDM-CIIR blends

Thermal properties	E80CI20	E80CI20NS2.5	E80CI20NS5	E80CI20NS7.5	E80CI20NS10
Initial degradation temperature (°C), T_i	202	205	210	232	203
Degradation temperature at the peak rate (°C), T_f	453	459	461	463	462
Terminal degradation temperature (°C), T_o	485	492	501	503	498
% char	92.6	94.9	95.1	94.7	94.2
Activation Energy (E_a), kJ/mol from HM equation	62.5	65.1	74.6	75.9	70.1
Activation Energy (E_a), kJ/mol from CR equation	24.4	25.4	25.6	27.0	22.7

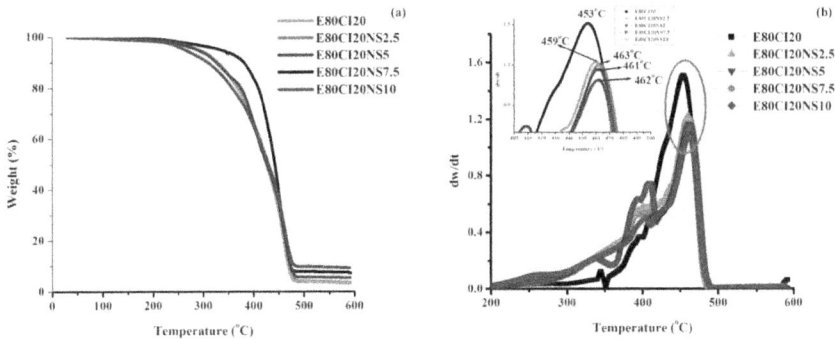

Figure 5.9 TGA and DTG thermograms of nanosilica reinforced EPDM-CIIR blends

5.2.6 (a) Kinetics of thermal degradation

The kinetics of thermal degradation predicts kinetic parameters which are related to thermal degradation mechanism. TGA provides a large amount of information for kinetic analysis and life time prediction under service conditions. The thermal decomposition kinetic analysis of the nanosilica reinforced EPDM/CIIR blends was investigated based on several model fitting methods. Coats-Redfern (CR) and Horowitz-Metzger (HM) models used in this study are based on TGA with constant heating rate [128]. CR equation is an integral method to evaluate activation energy (E_a) which involves thermal decomposition mechanism.HM equation also derived an approximate integral method to find E_a for first order reaction. According to HM equation, the slope of the plot of double logarithm of reactant weight fraction against temperature gives energy of activation as per the following equation

$$\ln\left[\ln\left[\frac{m_o - m_f}{m_t - m_f}\right]\right] = \frac{E_a\theta}{RT_m{}^2} \qquad (5.10.)$$

The reactant weight fraction or degree of conversion (α) is represented by $\alpha = \frac{m_o - m_f}{m_t - m_f}$, where m_o and m_f represents initial and final mass of sample, m_t is the mass associated at particular temperature. E_a represents the activation energy and $\theta = T - T_m$. T_m is the peak degradation temperature or temperature at which the degradation occurs at maximum rate and 'R' is the universal gas constant (R = 8.314J/molK). E_a (kJ/mol) can be calculated from the slope of the curve (= $E_a * 10^3 / RT_m{}^2$).

126

CR method is an integral method which includes different functions viz. Mampel power law, Avrami equation, Jander equation, Ginstling equation etc. on the basis of order of decomposition reactions. Coats and Redfern provided an approximation for first order reaction process. The simplified and integral form of rate equation is given as

$$\log \frac{g(\alpha)}{T^2} = \log \frac{AR(1-2RT/Ea)}{\Phi Ea} - \frac{Ea}{2.303RT} \qquad (5.11.)$$

Where T is the temperature in K, A is the pre-exponential term, R is the universal gas constant, Ea gives activation energy, Φ represents heating rate and α is given as $\alpha = \frac{mo-mt}{mo-mf}$, where m_o, m_t and m_f represents the initial, mass at temperature T and final masses respectively. $g(\alpha)$ is given by -log $(1-\alpha)$ from Mampel power law for first order kinetics (n=1). The values of $\frac{g(\alpha)}{T^2}$ and $\frac{1000}{T}$ are plotted against each other to give a straight line and energy of activation can be calculated from its slope $(\frac{Ea}{2.303R})$.

Upon reinforcement of EPDM/CIIR blend with nanosilica, the energy barrier required to overcome thermal agitation increases thereby increasing activation energy. The increase in energy of activation for thermal degradation seen in nanosilica filled EPDM/CIIR blends also reveals enhancement in thermal stability after reinforcement. The E_a increased by 21.4% for E80CI20NS7.5 when compared to unfilled blends as observed from Table 5.5.

5.2.7 Gamma irradiation aging effects of mSiO$_2$ reinforced EPDM-CIIR rubber blends

The mSiO$_2$ reinforced EPDM-Chlorobutyl rubber blends were exposed to cumulative doses (0.5MGy, 1 MGy and 2 MGy) of γ-radiations. The extent of crosslinking and/or chain scission effects also influences the properties. FTIR spectroscopy was utilized to evaluate the chemical changes produced after irradiation. The presences of free radicals generated upon irradiation in elastomers were evaluated by ESR analysis.

5.2.7 (a) FTIR spectra of γ-irradiated mSiO$_2$ reinforced EPDM- Chlorobutyl rubber blends

Attenuated total reflectance FTIR (ATR-FTIR) spectroscopy was carried out to determine the formation of new chemical bonds or functional groups after irradiation. FTIR analysis was carried out for all the mSiO$_2$ reinforced EPDM-Chlorobutyl rubber blends subjected to gamma rays exposure. For an illustration, FTIR spectra of E80CI20NS7.5 before and after irradiation are represented in Figure 5.10. The new peaks noticeable at 755 cm^{-1}, 971 cm^{-1}, 1032 cm^{-1}, 1087 cm^{-1}, 1375 cm^{-1} and 1538 cm^{-1} for EP80CI20NS7.5 subjected to 1 MGy radiation corresponds to C-C bonds. The formation of C-C bonds indicates the dominance of crosslinking in sample subjected to 1MGy. There was no noticeable change in FTIR spectrum subjected to 0.5MGy dose of gamma radiation. The FTIR spectra of E80CI20NS7.5 irradiated at 2MGy revealed the formation of new peaks corresponding to oxygenated bonds. The intense bands at 3309 cm^{-1}, 1136 cm^{-1} and 670cm^{-1} corresponds to –OH and –COO bonds. Thus dominance of chain scission or oxygenated degradation over crosslinking was observed in E80CI20NS7.5 subjected to 2 MGy. The free radicals generated after irradiation, in presence of molecular oxygen, react to form oxygenated groups (hydroxyl and peroxyl radicals). The compounds like carboxylic acids, hydroperoxides, ketones containing carbonyl and hydroxyl functional groups associates to oxygenated products. The hydroperoxide formed after reaction of peroxyl group with hydrogen atoms in rubber chains cleavages to form two new free radicals viz. hydroxyl and peroxyl radicals which further propagate and continue to cause degradation by chain mechanism [252].

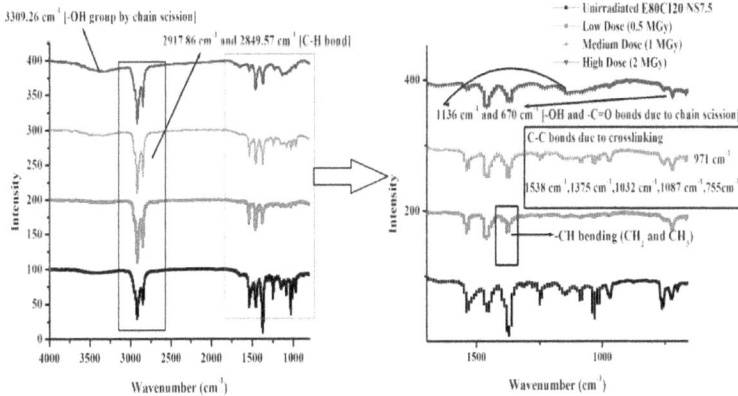

Figure 5.10. FTIR spectra of irradiated nanosilica reinforced EPDM-CIIR blend

5.2.7 (b) Electron Spin Resonance (ESR) spectroscopy

ESR spectroscopy is a reliable analysis for evaluating and quantifying the presence of free radicals in polymers. Elastomers generally do not possess paramagnetic centers and therefore

ESR study is not applicable. However, the radiation induced band gap excitation introduces paramagnetic centers in the elastomeric system, which gives possibility for application of ESR spectroscopy [253] [149]. ESR spectroscopy was carried out for irradiated and non-irradiated mSiO$_2$ reinforced EPDM-CIIR blends. As a representation, the ESR spectra of E80CI20NS7.5 (before and after radiation) are shown in Figure 5.11. The silane grafted nanosilica particles contain different types of oxygenated functional groups viz. hydroxyl and silanol groups. The increase in free radicals produced by irradiation reacts with such groups to form oxygenated radicals. ESR spectrum provides direct and accurate measurement on presence of free radicals (generated after irradiation) on the elastomeric surface. At 1MGy radiation exposure, no new peaks were formed for the nanocomposite implying that free radicals are not present in the structure. The free radicals or unpaired electrons generated after irradiation might have taken part in the formation of cross links [101], [250], [143]. In the case of nanocomposites irradiated at 2MGy, new peaks were observed corresponding to g values of 2.373, 1.167 and 1.378 signifying presence of free radicals [254]. These free radicals can induce chain scission which manifests in lowering of mechanical properties.

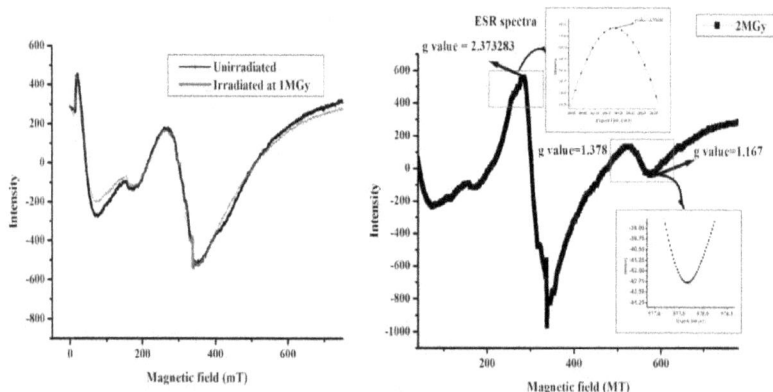

Figure 5.11 ESR spectra of irradiated nanosilica reinforced EPDM-CIIR blends

5.2.7 (c) Static mechanical properties of irradiated mSiO$_2$ reinforced EPDM-CIIR blends

The gamma radiation induced chain scission and/or cross-linking effects influence elastomer in two manners. The dominance of crosslinking increases mechanical tensile strength and modulus but reduces elongation at break. The increase of chain scission reduces tensile strength and modulus but increases elongation at break to some extent due to cleavage of elastomeric chains. A comparative study of tensile properties of nanosilica reinforced EPDM-Chlorobutyl rubber blends exposed to varying cumulative doses of γ-rays were studied in order to evaluate the influence of gamma irradiation on nanocomposites. The variation of tensile strength and its percentage change with irradiated neat blends for all gamma irradiated mSiO$_2$ reinforced EPDM-CIIR blends were documented in Table 5.6. The percentage change in modulus (M100) of all irradiated nanosilica reinforced blends is also tabulated in Table 5.6. It was evident that tensile strength and modulus for the nanosilica filled EPDM-CIIR blends enhanced till 1 MGy. Increase in tensile strength as a function of gamma irradiation can be primarily attributed to dominance in crosslinking amongst polymeric chains [112], [157], [255]. As seen from FTIR spectroscopy, the formation of new C-C bonds for sample irradiated at 1MGy also corresponds to formation of crosslinking sites. The decrease in solvent permeation till 1 MGy also attribute towards crosslinking. The reduction in mechanical properties for the sample irradiated at 2MGy was due to dominance of radiation induced chain scission or oxygenative degradation due to high dose of radiation [256]. The formation of oxygenated bonds and presence of free radicals as evident from FTIR and ESR analysis also converges towards oxygenative degradation or chain scission [143]. Amongst the nanosilica incorporated EPDM-CIIR blends, the nanocomposite with 7.5 phr nanosilica particles exhibited least change in mechanical properties. The well distributed surface grafted nanosilica particles in E80CI20NS7.5 as evident from TEM micrographs impart good radiation resistance. The covalent bonds created in between mSiO$_2$ and elastomer chains with higher bond dissociation energy imparted less radiation induced changes in the structure of nanocomposite. In addition, the presence of nanoparticles in the blend provided barrier to permeation of free radicals. The elastomeric materials should produce least change in properties when subjected to γ-irradiation in order to attain longevity for its utility in radiation environments [9], [257].

Table 5.6 Mechanical properties of irradiated nanosilica reinforced EPDM-CIIR blends

Nanosilica reinforced EPDM blends	Before radiation	Low dose (0.5 MGy)	Medium dose (1 MGy)	High dose (2 MGy)	Percentage change in TS on radiation exposure			Percentage change in M100 on radiation exposure		
					0.5MGy	1 MGy	2 MGy	(0.5 MGy)	(1 MGy)	(2 MGy)
E80CI20	1.38 ± 0.07	1.41±0.03	2.08±0.11	2.06±0.11	1.43	49.6	48.2	2.78	24.3	19.8
E80CI20 NS2.5	1.72 ± 0.05	1.77±0.03	2.18±0.07	2.07±0.08	2.90	26.7	20.3	3.52	18.5	19.8
E80CI20 NS5	1.98 ± 0.08	2.03±0.07	2.21±0.02	2.09±0.05	2.52	11.6	5.55	1.18	5.59	9.55
E80CI20 NS7.5	2.26 ± 0.11	2.31±0.06	2.41±0.04	2.15±0.02	2.21	6.63	-4.86	2.41	19.8	-7.58
E80CI20 NS10	1.59 ± 0.13	1.64±0.07	2.19±0.03	1.35±0.04	3.14	37.7	-15.1			-14.5

131

5.2.7 (d) Swelling behavior of γ-irradiated mSiO₂ reinforced EPDM-CIIR blends

The swelling behavior of γ-irradiated nanosilica reinforced EPDM blends were studied using cyclohexane as a solvent. The formation of radiation induced crosslinks till 1MGy reduced solvent permeation, as seen from FTIR and ESR spectroscopic analysis. Crosslinking restricts the mobility of elastomer chains and elastomer chain relaxation imparting barrier to solvent permeation. The dominance of radiation induced chain scission or cleavage of elastomer chains at higher dosage (2 MGy) provides ease of solvent permeation. The diffusion experiments were carried out for all irradiated samples. As an illustration, the irradiated blend with 7.5 phr mSiO₂ having well dispersed nanosilica particles, as discussed in preliminary sections of this chapter is demonstrated for sorption studies. The rate of solvent uptake has decreased drastically for the E80CI20NS7.5 irradiated at cumulative doses of 0.5 and 1MGy. As evident from mechanical, thermal and spectroscopic analysis, the dominance of crosslinking has resulted in less solvent uptake for samples irradiated at 1MGy.

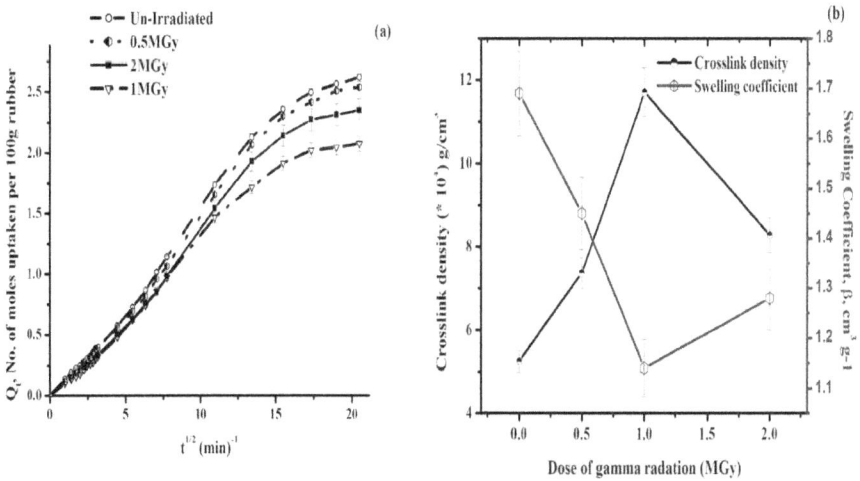

Figure 5.12 Sorption characteristics after irradiation of nanosilica reinforced EPDM-CIIR blends

The comparison of transport properties and percentage change in swelling coefficient as well as crosslink densities with neat blends are tabulated in Table 5.7 and 5.8 respectively. For representation, the transport coefficients of irradiated E80CI20NS7.5 tabulated. Similar observations were noticed for other nanosilica reinforced blends after irradiation.

Table 5.7 Transport coefficients of irradiated nanosilica reinforced EPDM-CIIR blend (E80CI20NS7.5)

Transport Coefficients				
	Low dose 0.5MGy	Medium dose 1.0MGy	High dose 2.0MGy	Un-irradiated
Crosslink density x 10^4 (gmol cc^{-1})	7.36	11.71	8.28	5.24
Diffusion coefficient (Dx10^7) (m^2 s^{-1})	3.12	2.59	2.85	3.35
Sorption coefficient (S) (g g^{-1})	1.86	1.46	1.64	2.16
Permeability coefficient (P x10^7) (m^2 s^{-1})	5.80	3.78	4.67	7.23
Swelling coefficient (β) (cm^3 g^{-1})	1.45	1.14	1.28	1.69

The significant increase in solvent uptake for the sample irradiated at 2MGy can be attributed towards radiation induced chain scission. The variation of crosslink density and swelling coefficient of E80CI20NS7.5 with respect to different cumulative doses of radiation are also plotted in Figure 5.12.

Table 5.8 Comparison of percentage change in transport coefficient after irradiation

Percentage change in swelling coefficient (β) at all doses of gamma radiation with unirradiated			
Sample	Low dose 0.5MGy	Medium dose 1.0MGy	High dose 2.0MGy
E80CI20	33.5	35.4	39.6
E80CI20NS2.5	32.9	34.9	39.5
E80CI20NS5	15.1	27.5	35.6
E80CI20NS7.5	14.2	22.5	34.2
E80CI20NS10	27.6	33.6	39.3

5.2.8 (e) Thermal degradation of γ-irradiated mSiO$_2$ reinforced EPDM-CIIR blends

The thermal degradation behavior of gamma irradiated mSiO$_2$ reinforced EPDM-CIIR blends was determined from TGA data. The initial, peak and final degradation temperatures of E80CI20NS7.5 irradiated at all doses are tabulated in Table 5.9. The increase in onset temperature of degradation at 1 MGy can be ascribed to dominance in radiation induced crosslinking. The decrease in onset degradation temperature at 2MGy is a characteristic of radiation induced chain scission. There was negligible difference in thermal properties for sample irradiated at 0.5MGy. These observations were in concurrence with mechanical and spectroscopic analysis. Similar trends were observed for all other mSiO$_2$ reinforced blends.

Table 5.9 Thermal properties of neat blend and nanosilica reinforced blend after irradiation

Thermal properties	E80CI20				E80CI20NS7.5			
	Before radiation	Low 0.5 MGy	Medium 1 MGy	Large 2 MGy	Before radiation	Low 0.5 MGy	Medium 1 MGy	Large 2 MGy
Initial degradation temperature (°C), T_i	202	210	225	223	232	233	238	229
Degradation temperature at the peak rate (°C), T_f	453	453	456	454	463	464	465	454
Terminal degradation temperature (°C), T_o	485	486	489	481	503	501	498	496
% char	92.6	92.9	95.4	90.9	94.3	94.4	94.5	96.2

5.3 Conclusions

Silane grafted nanosilica reinforced EPDM-CIIR blends were prepared with varying modified nanosilica, to investigate the enhancement in properties and gamma radiation behavior. The formation of covalent linkages after silane modification was directly identified from FTIR spectra. The changes in surface chemistry of nanosilica particles ensured dispersibility and interactions with the rubber chains. Morphological analysis confirmed evenly dispersed nanosilica upto 7.5 phr and formation of agglomerates at 10 phr. The dispersion and interfacial interactions of nanoparticles in the matrix enhanced static mechanical, viscoelastic and thermal characteristics while reducing solvent permeation. The suitability of theoretical models to predict sorption behavior and thermal degradation kinetic aspects were also investigated in this chapter.

The nanosilica reinforced blends were irradiated till 2 MGy, and least deviation in properties were observed at 7.5 phr nanosilica. Spectroscopic analysis confirmed chemical aspects of radiation induced effects on the nanocomposite.

Chapter 6

Studies on effect of MWCNT reinforcement on static mechanical, solvent sorption and gamma radiation ageing behavior of EPDM-CIIR blends[5]

6.1. Introduction

The carbon based nanoparticles like carbon nanotubes (CNTs), fullerene tubes, graphene, carbon nano-fibers (CNFs), carbon nano-horns (CNHs), carbon whiskers etc. has gained attention due to their outstanding mechanical and structural properties. Among carbon based nanofillers, multi-walled carbon nanotubes (MWCNTs) are considered very effective owing to its high mechanical strength and aspect ratio (>1000) [258], [259]. The strong covalent bond (σ bond) in the walls of MWCNT enables them to scavenge free radicals and absorb incident energy [93], [128]. Hence addition of MWCNTs to an elastomer can enhance radiation resistances while improving other physical properties like mechanical and transport properties. In this chapter, the influence of MWCNT filler on mechanical, solvent and radiation ageing characteristics of EPDM blends are investigated. The MWCNT based (0.5, 1, 1.5 and 2 phr) nanocomposites of EPDM-CIIR blends are prepared and evaluated for various properties.

6.2 Results and discussions

6.2.1 FTIR Analysis

The FTIR spectra of the EPDM-CIIR nanocomposites were analyzed to understand the occurrence of chemical interactions between the rubber and MWCNT [189]. For representation, the FTIR spectra of E80CI20, MWCNT and E80CI20CNT1.5 are plotted in Figure 6.1. A sharp peak at 1633 cm^{-1} corresponding to the –COOH group was observed in treated MWCNT. The FTIR spectra of the nanocomposites were compared with those of unfilled blend and MWCNT. In nanocomposites the intensity at this wave number was not significant. The –COOH groups in

---

[5] This chapter has been published in *Journal of Applied Polymer Science* (Neelesh Ashok *et al.*, "The influence of MWCNT and hybrid (MWCNT/nanoclay) fillers on performance of EPDM-CIIR blends in nuclear applications: Mechanical, hydrocarbon transport and gamma – radiation ageing characteristics", *2020*, 137 (42): 49271).

MWCNT reacted with rubber creating chemical interaction between the filler and the matrix. The FTIR spectra of nanocomposites are characterized by formation of a new peak at 1538 cm^{-1} and increase in peak intensity at 1646 cm^{-1} and 1046 cm^{-1}. It has been reported in literature [260] [261], [262] that incorporating MWCNT in a polymer matrix results in the formation of a secondary covalent network. In the nanocomposites, the presence of secondary covalent network is evident from the increase in peak intensity at 1646 cm^{-1} corresponding to C=C aromatic bond stretching of the MWCNT wall [260]. These bonds create anchoring sites with polymer chains resulting in improved mechanical properties [263] as explained in the coming sections. The rise in peak intensity at 1046 cm^{-1} when compared to unfilled blend (E80CI20) can be attributed to formation of carbonyl groups (–C=O) resulting from chemical interaction between –COOH group in MWCNT and rubber blend. The electron withdrawing chlorine atoms in chlorobutyl rubber react with carboxylic acid of MWCNT [264]. Due to difference in polarity, the electronegative chlorine atoms react with positive hydrogen atom in carboxyl group in MWCNT to form pi covalent carbonyl bond in the nanocomposites. In unfilled blend the peak at wave number of 1460 cm^{-1} indicate bending of –CH$_2$. Thus it may be concluded that FTIR spectra provided evidence on chemical interactions of -COOH modified MWCNT [265] with the EPDM-CIIR blend.

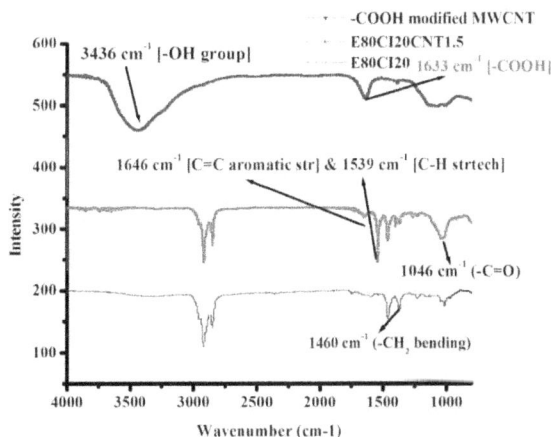

Figure 6.1 FTIR spectra of blend, -COOH modified MWCNT and nanocomposite

138

6.2.2 Cure characteristics

The cure characteristics are useful in analyzing various parameters such as processability, optimum cure time and physical properties of the polymer. The optimum cure time, cure rate index and scorch time (t_{s2}) are evaluated and analyzed from oscillating disc rheometre. The cure characteristics of MWCNT reinforced EPDM-CIIR blends are tabulated in Table 6.1. The t_{s2} and t_{90} of the MWCNT filled blends are higher than unfilled blends. Scorch time is a measure of the onset of curing and it may be noted that absorption of curatives by functional groups like – COOH, -OH, -C=O on to the MWCNT walls increases ts_2 [266]. With increase in MWCNT content, the t_{90} or optimum cure time, also increased. Increase in MWCNT content increases the restrictions in curative-matrix interaction and hinders the formation of cross-link sites [262], [265]. Hence, nanocomposites based on MWCNT require more time to cure than unfilled blends. For the same reason, other cure characteristics like CRI and $\Delta\tau$ (difference between τ_{max} and τ_{min}) which are measures of the extent of cross linking were also observed to decrease with increase in MWCNT content [267]. The enhancement in rubber-filler interactions restricts the mobility of elastomer chains thus requiring extra resistance for straining [186], [268].

Table 6.1 Cure characteristics of MWCNT reinforced EPDM-CIIR blends

Sample	Scorch time t_{s2}(min)	Optimum cure time t_{90} (min)	CRI (min^{-1})	Minimum Torque τ_{min} (N-m)	Maximum Torque τ_{max} (N-m)	(τ_{max}-τ_{min}) (N-m)
E80CI20	1.20	7.55	16.1	2.45	4.67	2.22
E80CI20 CNT0.5	1.47	6.23	21.4	11.8	50.4	38.6
E80CI20 CNT 1.0	1.55	6.32	20.6	12.2	45.1	32.9
E80CI20 CNT 1.5	1.72	6.75	19.8	12.9	44.6	31.7

E80CI20 CNT 2.0	1.35	6.73	18.6	13.3	42.4	29.1

6.2.3 TEM analysis

TEM micrographs of MWCNT reinforced EPDM-CIIR blends are shown in Figures 6.2 (a-d). The effectiveness of nanofiller in enhancing polymer properties depends on its state of dispersion in the matrix. The dispersion and extend of distribution of MWCNT in EPDM – CIIR blends were assessed using TEM. The TEM micrographs of E80CI20CNT0.5, E80CI20CNT1, E80CI20CNT1.5 and E80CI20CNT2 are shown in Figure 6.2 (a), (b), (c) and (d) respectively. TEM images of E80CI20CNT0.5, E80CI20CNT1 and E80CI20CNT1.5 reveal good dispersion of MWCNT. As content of MWCNT increased in E80CI20CNT2, as shown in Figure 6.2 (d), agglomeration was observed.

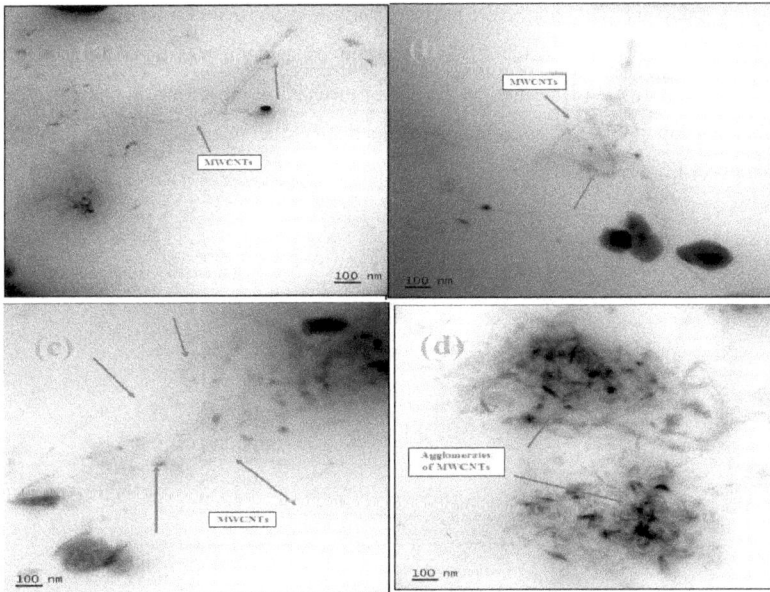

Figure 6.2 TEM micrographs of (a) 0.5 (b) 1 (c) 1.5 (d) 2 phr MWCNT in EPDM-CIIR blends

6.2.4 Mechanical properties

The static mechanical properties and stress-strain plots of MWCNT reinforced EPDM-CIIR blends are presented in Table 6.2 and Figure 6.3 respectively. The tensile strength (TS), and modulus (M300) of MWCNT reinforced EPDM/CIIR blends increased upto 1.5 phr MWCNT. A slight decline in tensile strength and M300 were observed for E80CI20CNT2 due to the formation of aggregates. The increase in mechanical properties was highest for 1.5 phr MWCNT with 69% increase in TS. The evenly dispersed MWCNT provide large interface with the polymer chains. The entanglement of polymer chains on the filler and the interfacial polymer – MWCNT interactions restricted the chain mobility [258], [263]. Larger interface also permit applied stress to be effectively transferred to the MWCNT [93], [120], [122]. Both these aspects contribute to improved reinforcement and mechanical properties in MWCNT reinforced blends [259]. Elongation at break increased for MWCNT reinforced blends when compared with neat ones because MWCNTs are flexible fillers which provide more ductility to the polymer nanocomposite without compromising its rigidity [269]. Agglomerates cause non-uniform filler distribution, reduction in polymer-filler interface and lower physico-chemical interactions [114], [270], [271]. Hence in EP80CI20CNT2, formation of agglomerates as evident from TEM micrograph in Figure 6.2 (d) resulted in decline of mechanical properties. In addition, stacks of nanotubes develop more stress concentration in the matrix-filler interfaces limiting the efficiency in load transfer from rubber matrix to fillers [117], [215].

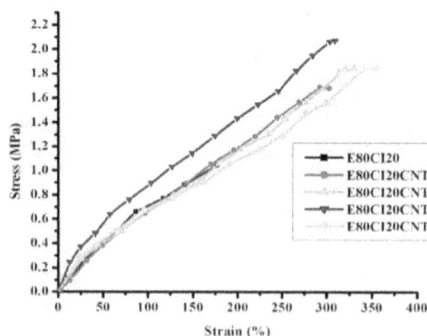

Figure 6.3 Stress strain plot of MWCNT reinforced EPDM-CIIR blends

Table 6.2 Tensile mechanical properties of MWCNT reinforced blends

Sample	Tensile strength, MPa	Tensile modulus (M300), MPa	Elongation at break, E_b (%)	Percentage change in TS
E80CI20	1.35 ± 0.07		97± 3	
E80CI20CNT0.5	1.70 ± 0.08	1.48±0.05	303±4	25.9
E80CI20CNT1	1.95 ± 0.09	1.53±0.06	321±3	44.4
E80CI20CNT1.5	2.28 ± 0.06	1.89±0.15	333±2	68.9
E80CI20CNT2	1.92 ± 0.08	1.70±0.08	342±4	42.2

6.2.5 Solvent sorption studies

Solvent sorption behavior of elastomer nanocomposites is a function of several factors like polymer segmental mobility, distribution and orientation of fillers etc. The study of sorption properties of the nanocomposites also serves as an effective tool to get an insight about morphology and filler-polymer interfacial interactions [262], [265], [272], [273]. The swelling of an elastomer is influenced by solvent transport characteristics. The transport coefficients (Permeation, Diffusion and Sorption coefficients) of MWCNT based nanocomposites of EPDM-CIIR blends are evaluated from sorption isotherms plotted in Figure 6.4. The solvent uptake for MWCNT and reinforced EPDM-CIIR blends was significantly lesser than unfilled ones [117], [273]–[275]. The calculated values of the transport coefficients are presented in Table 6.3.

Table 6.3 Solvent coefficients of MWCNT reinforced EPDM-CIIR blends

Sample	Diffusion coefficient $(Dx10^7)$ m^2 s^{-1}	Sorption coefficient (S) (g g^{-1})	Permeability coefficient (P $x10^7$) (m^2 s^{-1})	Swelling coefficient (β) (cm^3 g^{-1})
E80CI20	3.51	1.59	5.58	2.04
E80CI20CNT0.5	3.49	1.53	5.33	1.97
E80CI20CNT1	3.41	1.49	5.09	1.85
E80CI20CNT1.5	3.35	1.36	4.56	1.53
E80CI20CNT2	3.23	1.29	4.17	1.36

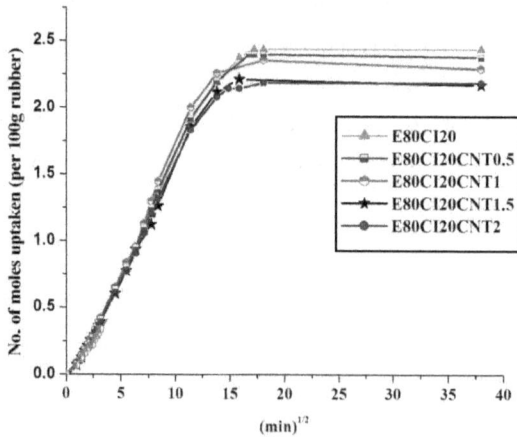

Figure 6.4 Sorption isotherms of MWCNT reinforced EPDM-CIIR blends

It can be observed that the swelling, diffusion, sorption and permeation coefficients for the MWCNT reinforced blends were lower than unfilled blends [273], [274]. The reinforcement of blend with MWCNT reduces the free space as discussed in the previous chapters and reduces the solvent permeation [266], [276].

6.2.6 γ-Irradiation studies

The MWCNT reinforced EPDM-CIIR blends were exposed to ^{60}Co- γ radiation and effect of irradiation for different cumulative doses were studied in this section. The gamma irradiation of elastomer chains generates free radicals (unpaired electrons) that induce either cross-linking and/or degradation of chains, depending on the dosage of radiation. The cross-linking and chain scission effects were analyzed from the FTIR spectra of the nanocomposites exposed to γ radiation. Electron Spin Resonance spectroscopic analysis was also carried out to detect the presence of free radicals [254], [268]. The effect of radiation dose on mechanical and transport properties of the nanocomposites were also evaluated and discussed below.

6.2.6 (a) FTIR analysis of γ-irradiated MWCNT reinforced EPDM-CIIR blends

The FTIR spectra of all irradiated MWCNT reinforced blends were analyzed to study the chemical interactions after exposing to γ radiation. For representation, FTIR spectra of irradiated and non-irradiated E80CI20CNT1.5 are depicted in Figure 6.5. No significant changes in the spectra were observed for samples exposed to low dose (0.5 MGy) of gamma radiation. In nanocomposites exposed to 1MGy (medium dose) radiation, the formation of absorption peaks at 1200-1300 cm^{-1}range (C-C bonds-stretching) resulted from radiation induced crosslinking. For the nanocomposites exposed to 2MGy (highest dose) of gamma radiation, intense bands were observed at wavenumbers 1031 cm^{-1},1047 cm^{-1} and 1746 cm^{-1} corresponding to C-O (stretch) C-O-C (asymmetric stretch) and C=O respectively [260]. These oxygenated bonds are formed due to the dominance of chain scission or degradation of the polymer chains. The cleavage of bonds gives rise to free radicals which react with atmospheric oxygen to form peroxyl radicals. The functional groups like carbonyl, ketones, hydroperoxides and hydroxyl arise from oxygenated compounds formed by the reaction of peroxyl radicals. The intensity of peak at 1646 cm^{-1} decreased at 2 MGy. At higher radiation dose, chain scission is more dominating than crosslinking.

6.2.6 (b) ESR spectroscopy of γ-irradiated MWCNT reinforced EPDM-CIIR blends

Electron spin resonance (ESR) spectroscopy is a reliable and direct measurement technique to identify and evaluate the free radicals formed in the polymer on exposure to radiation. The radiation induced changes like crosslinking and/or chain scission in elastomeric materials generates free radicals.

ESR spectra of unirradiated and irradiated MWCNT reinforced EPDM-CIIR blends are plotted in Figure 6.6. For medium radiation dose, there was no significant change in spectra. The free radicals generated upon radiation took part in cross-linking reactions as evident from the FTIR spectra. It is also reported in literature that MWCNT act as radical scavenger [254], [268], [277]. At high dose (2 MGy) some free radicals were observed corresponding to g value (2.31-2.35 range) in the ESR image [253], [278], [279]. For high dose of radiation (2MGy), the g value in ESR spectra arise from the presence of carbonyl free radicals. This observation is in concurrence with the analysis of FTIR spectra of the nanocomposites.

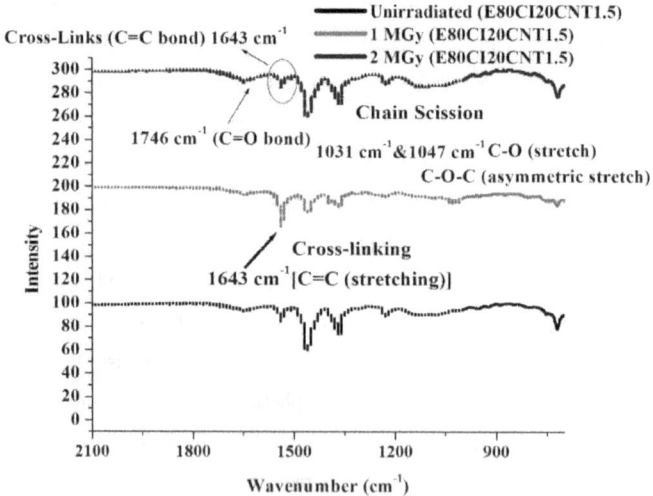

Figure 6.5 FTIR spectra of irradiated MWCNT reinforced EPDM-CIIR blend

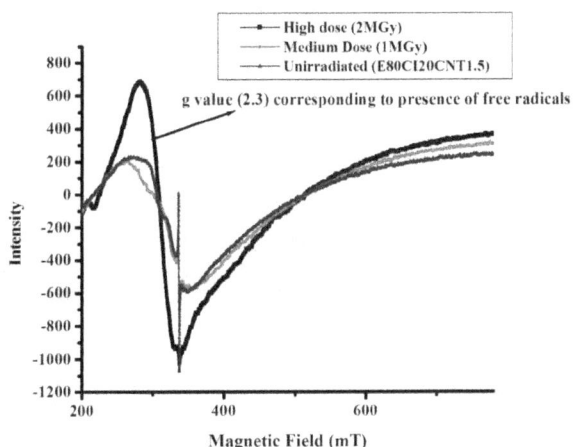

Figure 6.6 ESR spectra of irradiated MWCNT reinforced EPDM-CIIR blend

6.2.6 (c) Static mechanical behavior of γ-irradiated MWCNT based EPDM-CIIR blends

The mechanical properties of irradiated nanocomposites are tabulated in Tables 6.4 and 6.5 shown below. The percentage change in mechanical properties was calculated using the equation given in Chapter 3.

Table 6.4 Mechanical tensile strength in gamma-irradiated MWCNT based EPDM-CIIR blends

EPDM-CIIR Nanocomposite	Before Irradiation	Low Dose 0.5MGy	Medium Dose 1MGy	High Dose 2MGy	Percentage change in TS on radiation exposure		
					0.5 MGy	1 MGy	2 MGy
E80CI20	1.37±0.07	1.39±0.08	2.08±0.11	2.06±0.07	1.45	51.8	50.3
E80CI20CNT0.5	1.70±0.08	1.76±0.03	2.18±0.07	2.01±0.09	3.52	28.2	18.2
E80CI20CNT1	1.95±0.09	2.03±0.07	2.21±0.02	2.02±0.06	4.10	13.3	3.58
E80CI20CNT1.5	2.28±0.06	2.31±0.06	2.38±0.04	2.26±0.03	1.31	4.38	-0.87
E80CI20CNT2	1.92±0.08	1.95±0.03	2.04±0.02	2.03±0.04	1.56	6.25	5.72

It was observed that, in nanocomposites exposed to gamma irradiation of upto 1 MGy, tensile strength and M100 increased. The increase in mechanical properties is a consequence of the crosslinking of polymer chains upon irradiation, which is evident from the FTIR spectra. At higher dose of 2 MGy, there was a decline in mechanical properties owing to the prevalence of degradation or chain scission. MWCNT scavenges the free radicals generated during irradiation which reduces the damages caused to the elastomeric structure [268], [277], [280]. The least change in mechanical tensile strength and modulus were observed for E80CI20CNT1.5. The well distributed MWCNTs in 1.5 phr content as evident from TEM micrographs given in Figure 6.2, provides more radiation resistance to the EPDM-CIIR blends.

Table 6.5 Modulus of gamma-irradiated MWCNT based EPDM-CIIR blends

MWCNT reinforced EPDM-CIIR blends	M300 (MPa)				Percentage change of M300 with un-irradiated		
	Un-irradiated	Low dose 0.5MGy	Medium dose 1MGy	High dose 2MGy	Low dose 0.5MGy	Medium dose 1MGy	High dose 2MGy
E80CI20CNT0.5	0.78	0.96	1.49	0.76	23.0	91.0	-2.56
E80CI20CNT1	1.23	1.28	1.53	0.78	4.06	24.4	-36.6
E80CI20CNT1.5	1.89	1.88	1.92	0.80	-0.52	1.58	-55.3
E80CI20CNT2	1.87	1.88	1.91	0.79	0.53	2.13	-57.7

6.2.6 (d) Transport Properties

Transport properties for MWCNT nanocomposites of EPDM-CIIR blends are tabulated in Table 6.6. The transport coefficients for all MWCNT based composites of EPDM blends were evaluated, as a representation the properties of E80CI20CNT1.5 were tabulated. Sorption coefficients decreased after exposure to gamma radiation due to formation of radiation induced crosslinking. In this case, the presence of MWCNT imparts better barrier properties after irradiation for the nanocomposite compared with neat blends. It can be inferred that the presence of nanotubes offers more tortuous path for the solvent to permeate through the degraded nanocomposite [281]. For high dose (2 MGy), dominant chain scission decreased the crosslink density and hence all the transport coefficients increased. MWCNT reinforced EPDM-CIIR blends had better hydrocarbon solvent barrier properties as compared to unreinforced blends when subjected to γ-radiation.

Table 6.6 Transport properties of irradiated MWCNT reinforced blends

Transport Coefficients	E80CI20CNT1.5				Percentage change in transport coefficients of MWCNT nanocomposites on radiation exposure		
	Before	Low dose 0.5 MGy	Medium dose 1 MGy	High dose 2 MGy	Low dose 0.5 MGy	Medium dose 1 MGy	High dose 2 MGy
Diffusion coefficient $(Dx10^7)$ m^2 s^{-1}	3.35	3.12	2.59	4.30	-6.86	-22.68	28.35
Sorption coefficient (S) $(g\ g^{-1})$	1.36	1.32	1.24	1.42	-2.94	-8.82	4.41
Permeability coefficient $(P\ x10^7)$ $(m^2\ s^{-1})$	4.56	4.12	3.21	6.11	-9.64	-29.6	33.9
Swelling coefficient (β) $(cm^3\ g^{-1})$	1.53	1.49	1.37	1.82	-2.61	-10.4	18.9

6.3 Conclusions

MWCNT reinforced nanocomposites of EPDM-CIIR blends were prepared by a two-stage solid state mixing. The prepared nanocomposites were evaluated for cure characteristics, mechanical and solvent sorption behavior. FTIR analysis confirmed the presence of chemical interactions between the filler and blend matrix. The dispersion of MWCNTs was analyzed by TEM. The interfacial interactions and effective stress transfer between the nanofiller and matrix resulted in enhancement of mechanical properties with increasing nanofiller content till 1.5 phr MWCNT. Solvent transport coefficients were found to decrease with filler content owing to induction of tortuous path and reduction in free volume by the nano reinforcements. The slight reduction in mechanical properties of blends at higher content of MWCNT may be attributed to agglomeration. MWCNT reinforced nanocomposites of EPDM-CIIR blends were exposed to cumulative doses [0.5MGy, 1MGy and 2 MGy] of γ-radiation. Chemical interactions and presence of free radicals upon irradiation were analyzed by FTIR and ESR spectroscopy. Up to medium dosage of 1 MGy, the crosslinking effect was dominant while for higher dosage of 2 MGy, chain scission was predominant. MWCNT act as free radical scavenger which when reinforced in blends produced minimized damage caused due to γ-radiation when compared with neat blends. The free radical scavenging effect of MWCNT along with the induction of tortuous path for free radical migration imparted superior radiation resistance to EPDM-CIIR blends. MWCNT reinforced EPDM-CIIR blends have the potential for application in rubber components that are used in radiation and hydrocarbon ageing environment.

Chapter 7

Studies on carbon black-nanofiller hybrid composites of EPDM-CIIR rubber blend for product application[6]

7.1 Introduction

In this chapter, the hybrid nanocomposites of EPDM-CIIR blends based on carbon black and nanofillers were prepared and evaluated for mechanical properties, solvent sorption characteristics and gamma radiation ageing behavior. The preliminary studies on nanocomposites of EPDM-CIIR blends based on three types of nanofillers with different geometries were carried out in earlier chapters. For product application in nuclear fuel reprocessing facilities, the properties of rubber can be further improved by reinforcing with fillers like carbon black [81], [282]. Generally, carbon black and silica are conventionally used reinforcing agents for achieving good strength and stiffness required for prescribed applications in rubber compound manufacturing since decades [95], [283]–[285]. The reinforcing efficiency of elastomers with carbon black can be remarkably enhanced by incorporating nanoscale fillers [103], [208], [283]. Carbon black-nanofiller hybrid composites [123], [175], [284] of elastomers have attained considerable attention for functioning in intense ageing environments due to their improved properties [121], [283]. The improvement in properties is based on synergistic effect of hybrid fillers [123], [286], preparation methods and nanofiller-rubber interfacial interactions.

Zainathul *et al.* [175] reported a detailed review on synergistic effect of hybrid fillers (carbon black or silica as primary fillers and secondary fillers from various sources) on cured rubbers . Choi and coworkers [287], [288] focussed on evaluating cure characteristics, morphology and mechanical properties of hybrid filler-rubber composites. In addition, several researchers have investigated on the improvement in filler-rubber interactions of carbon black/nanoclay and nanosilica reinforced elastomers [260], [289]. Asish Malas and Chapal Kumar Das [283] explored the thermal and mechanical properties of EPDM/carbon black rubber composites

---

[6] This chapter has been published in *Materials Research Express* (Neelesh Ashok and Meera Balachandran, "Effect of nanoclay and nanosilica on carbon black reinforced EPDM/CIIR blends for nuclear applications", *2019*, 6(12): 125364.

reinforced with three types of organomodifed nanoclay (Cloisite 15A, Cloisite 20A and Cloisite 30B). Micro and nanosized silica (SiO_2) [110], [290] particles with carbon black are also effective reinforcements for rubbers in several applications. Peter and coworkers [291] have investigated the mapping of surface elastic moduli in EPDM rubbers reinforced with nanosilica particles using AFM. Haisheng and Avraam [289] carried out comparative study on rheological, dynamic mechanical and vulcanization properties of silica, nanoclay and CB filled EPDM rubber.

The elastomeric materials used in nuclear fields are exposed to different corrosive chemicals like paraffinic hydrocarbon solvents, nitric acid, etc. apart from gamma rays as stated in introduction chapter [6], [20], [47]. They not only have to withstand radiation but also prevent permeation of hydrocarbons into them. The fissionable material is recovered from spent nuclear fuel by extraction process that uses solvents like alkyl hydrocarbons and tributyl phosphate as well as nitric acid. Though, CNTs are good radiation resistant reinforcing fillers for polymers [121], [128], the ability to withstand in long chain hydrocarbon solvents and nitric acid is remarkably low as the carbon based nanomaterials react with solvents leading to breakage of fillers [292]–[294].

In this context, hybrid nanocomposites are prepared by reinforcing EPDM/CIIR blends with varying amounts of organo modified nanoclay and silane modified nanosilica along with carbon black. The synergistic effect of hybrid fillers and filler-rubber interactions in enhancing the properties of EPDM-CIIR blends are discussed in this chapter.

7.2 Results and discussions

7.2.1 FTIR analysis

FTIR spectra of nanoclay and nanosilica reinforced EPDM-CIIR blends are shown in Figure 7.1. FTIR spectra of hybrid filler reinforced EPDM/CIIR blends were analyzed to estimate the occurrence of chemical interactions between nanofillers and blend. For illustration, FTIR spectra of EP80CI20NC5 and EP80CI20NS5 were compared with EP80CI20 and are shown in Figure 7.1. New peaks observed at 1540 cm^{-1} and 1015 cm^{-1} corresponding to –NO and –CN bond arise from chemical reaction between organomodifer group in nanoclay and blend. The increase in intensity of peaks at 720 cm^{-1} for both NC and NS reinforced EPDM/CIIR blends corresponds to

151

methylene group (-CH₃). The presence of –CH₃ in the rubber denotes more regularity in the back bone structure of the composite (long range linearity of elastomer chains). In FTIR spectra of EP80CI20NS5, new peak seen at 1029 cm^{-1} were attributed to formation of Si-O-Si linkage which is characteristic of surface functionalized nanosilica. The reduced intensity at 1454 cm^{-1} wavenumber showed decline in flexural vibration of –CH bending (scissoring). FTIR spectra provided information on occurrence of chemical interactions when nanofillers are incorporated in the matrix. The presence of CB in all the samples is indicated by presence of common peak at 1640 cm^{-1} and 1700 cm^{-1}, the region corresponding to C=O conjugated bonds present in carbon black. Few peaks observed in 3400 cm^{-1}-3600 cm^{-1} wavenumber region is attributed to –OH group present in carboxylic acid functional group of CB.

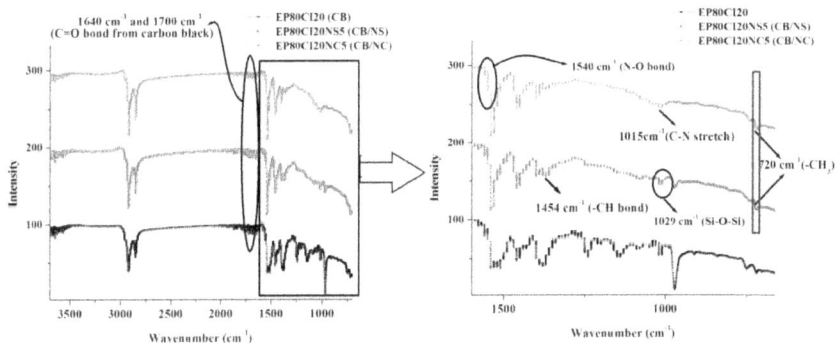

Figure 7.1 FTIR spectra of hybrid nanocomposites of EPDM-CIIR blends

7.2.2 Morphology

The degree of dispersion of nanofillers (NC and NS) in the CB reinforced EPDM/CIIR blend determines the enhancement in properties of nanocomposite. The variation in position, intensity and breadth of the XRD spectra provides information on the structure of NC as well as state of dispersion and exfoliation of NC in the rubber matrix. The layered structure in nanocomposite gives rise to a peak in the XRD spectra. The XRD spectra of nanocomposites (with and without nanofillers) are plotted in Figure 7.2. For nanoclay used in this work, the peak appeared at $2\theta=4.34°$ as shown in Figure 7.2 and the d spacing was calculated to be 20.15A° (2.015 nm). Compared to nanoclay, the spectra of the EP80CI20NC5NS0 and EP80CI20NC10NS0 do not show any prominent peaks. This implies that the layers of the nanoclay could have delaminated and exfoliated between the EPDM/CIIR rubber chains. However, the shift in peak observed at $2\theta = 6.49°$ and $2\theta = 6.43°$ for EP80CI20NC5NS0 and EP80CI20NC10NS0 implies that there was a decline in d-spacing of NC. The d-spacing of NC in the binary nanocomposites was calculated to be 1.36 nm. It has been postulated that alkyl group in organo-modified clay participates in curing reaction resulting in collapse of layers of nanoclay and decrease in d-spacing as detailed in Chapter 4. However, a closer look at spectra, it shows smaller peaks of lower intensity near 4.3° for EP80CI20NC10NS0. The smaller peaks arise from the presence of agglomerates of NC that retains in the layered structure. There were no identifiable peaks for EPDM-CIIR blends reinforced with CB. From XRD results, it can be inferred that organo-modified layered silicates (nanoclay) are well dispersed in EPDM-CIIR blends with exfoliation and slight agglomeration at 10 phr.

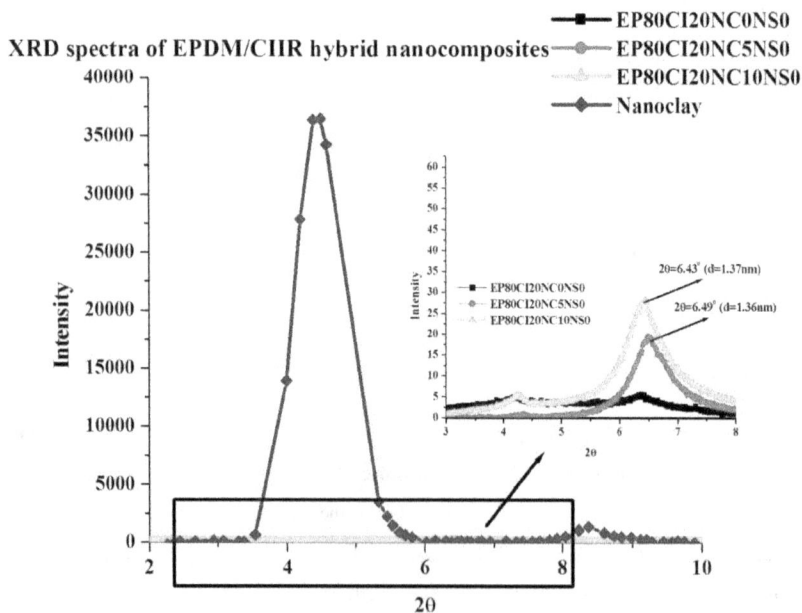

Figure 7.2 XRD spectra of hybrid nanocomposites of EPDM-CIIR blends

TEM provides qualitative information about morphology of hybrid nanocomposite through direct visualization. The efficiency of nano-reinforcement in rubbers is dependent on the state of dispersion of nanofillers in the blend. TEM micrographs of CB/nanosilica and CB/nanoclay reinforced EPDM-CIIR blends are shown in Figure 7.3(a) and (b). In TEM micrograph of EP80CI20NC5 (Figure 7.3(a)), well dispersed platelets of layered silicates can be observed. Few stacks of nanoclay were observed in TEM micrograph of EP80CI20NC10 (Figure 7.3(b)).

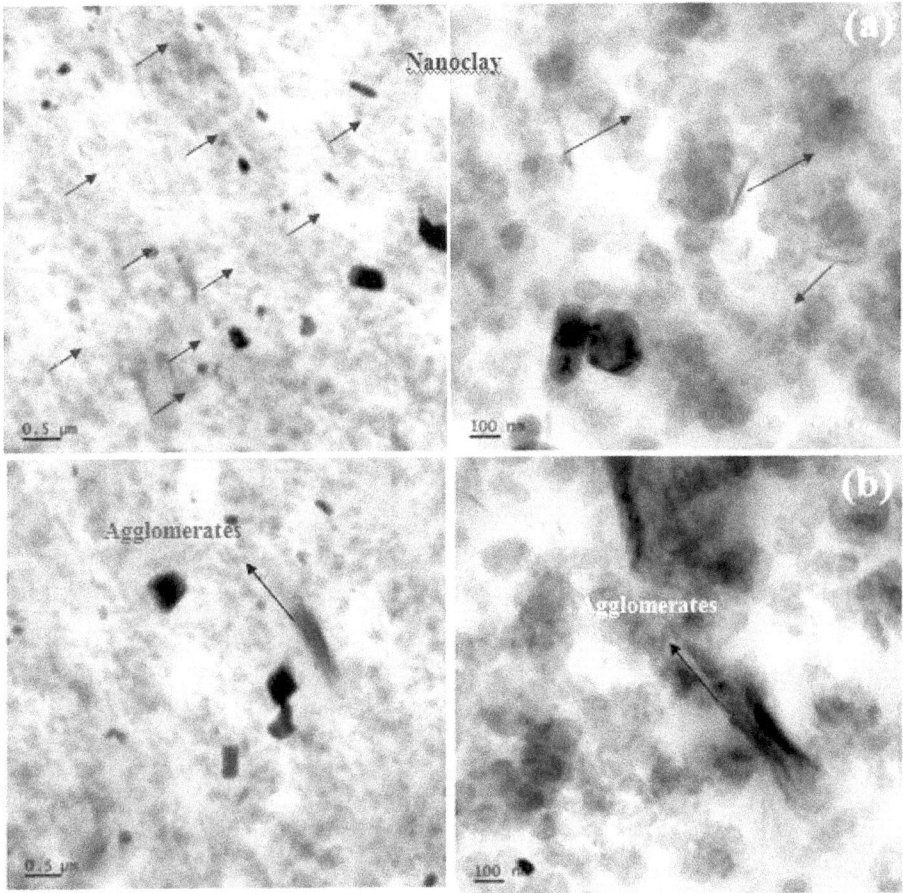

Figure 7.3 TEM micrographs of (a) EP80CI20NC5 and (b) EP80CI20NC10

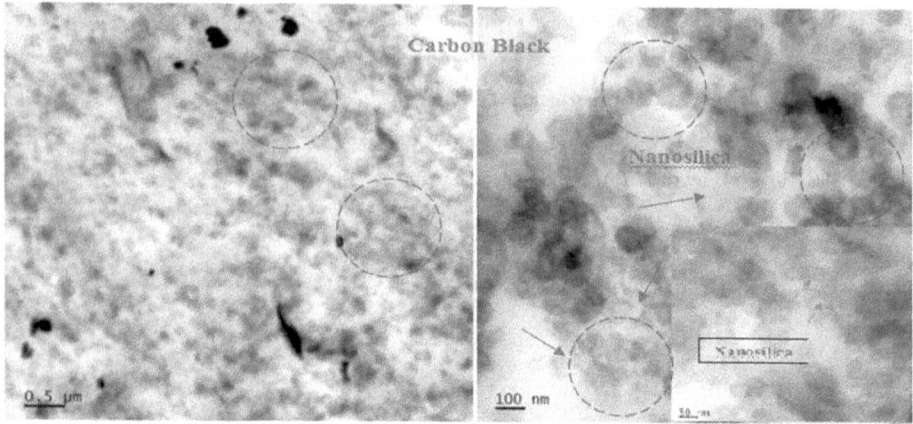

Figure 7.4 TEM micrographs of EP80CI20NS5

7.2.3 Payne effect

The effect of the hybrid reinforcement can be understood by the deformation induced changes in the polymer/filler network. The results obtained from strain-sweep measurements demonstrate the features attributing to Payne effect. The fundamental reinforcement mechanism and non-linear viscoelastic behavior was because of interactions between entangled rubber chains and filler particles as discussed in earlier chapters. Payne effect was studied on all compositions of hybrid fillers. As a representation, the effect of strain sweep on shear storage modulus (G') on nanoclay reinforced EPDM-CIIR blends are shown in Figure 7.5(a). It can be observed that G' increased with NC content. Sudden drop in G' in EP80CI20NC10 is due to more filler-filler interactions which tend to break easily at higher strains. But in EP80CI20NC5, a slow and uniform decline in G' was observed due to stronger filler-rubber interfacial interactions. Large specific surface area of nanofiller create larger interface with the elastomer chains. In EP80CI20NC5, more filler-rubber interactions were generated as a result of uniform dispersion and well exfoliation of layered silicates in the CB filled blend as evident from TEM micrograph. The similar trend in Payne effect was observed for NS reinforced EPDM/CIIR blends. In case of hybrid composites with combined fillers (EP80CI20NC5NS5 and EP80CI20NC5NS10), G' is generally higher than other specimens due to synergistic effect of both fillers. But, at higher filler

155

content for composites (with both NC and NS content) the drop in G' was steep and sudden indicating more filler-filler interactions as shown in Figure 7.5(b). The larger nanofiller-rubber interactions along with synergistic effect of fillers as schematically represented in Figure 7.6 with different geometries established enhancement in mechanical properties measured in static and shear mode.

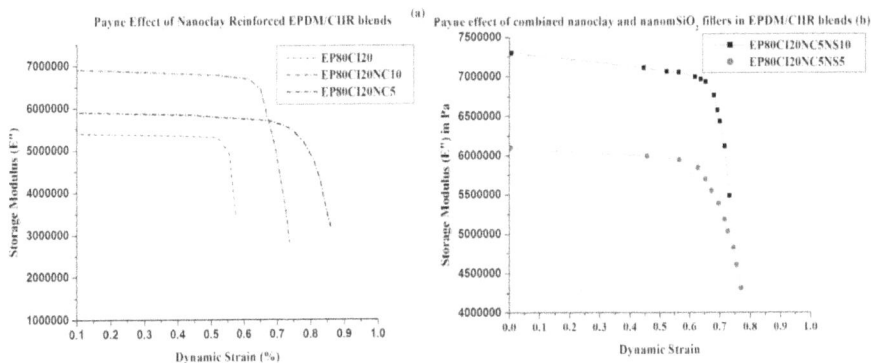

Figure 7.5 Payne effect of (a) nanoclay and (b) hybrid filler based blends

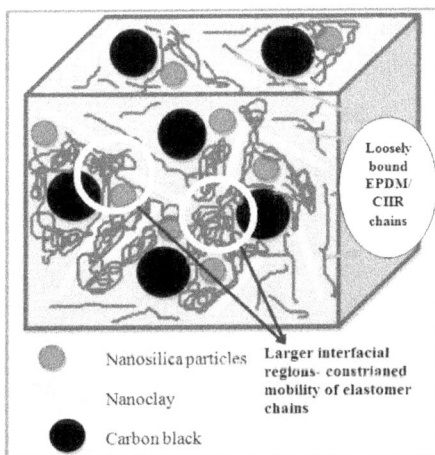

Figure 7.6 Schematic representation of hybrid fillers in EPDM/CIIR blends

7.2.4 Static mechanical properties

The static mechanical properties of hybrid filler reinforced EPDM/CIIR blends are tabulated in Table 7.1. The tensile strength and modulus (M100) at 100% elongation increased with nanofiller content. The evenly dispersed NC and NS in CB reinforced EPDM/CIIR blends provide larger interface with the elastomer chains. The interfacial rubber-nanofiller interactions and entanglement of elastomer chains on the filler restricted chain mobility as represented in Figure 7.6. The increase in tensile strength and M100 is higher for NC reinforced blends compared to NS incorporated ones. The reason can be attributed to the fact that rigid filler like layered silicate or nanoclay takes up major portion of the force or stress applied on the EPDM/CIIR hybrid nanocomposite and prevents stress transfer to elastomer chains. This eliminates the cleavage of elastomeric chains while straining resulting in property enhancement. The layered or planar structure of nanoclay imparts greater interface to EPDM/CIIR matrix than spherical silica particles. The evenly distributed nanoSiO$_2$ particles in EPDM/CIIR blends also established larger interfaces with polymer chains, resulting in enhancement in mechanical properties. From Table 7.1, it can also be observed that swelling coefficient (β) also decreased with increase in NC and NS content. The least value of β was observed for EP80CI20NC10NS5. It is due to the fact that dispersed NC and silica particles in EPDM/CIIR blend imparted more "tortuosity" for the solvent permeation. The mechanical properties (tensile strength and M100) are higher for samples reinforced with both NC and NS due to synergism of combined fillers with CB.

Table 7.1 Tensile properties of hybrid nanocomposites of EPDM/CIIR blends

Nanoclay	Nanosilica	Tensile strength (MPa)	Elongation at break, E_b	M100 (MPa)	Swelling coefficient, β (cm^3/g)
0	0	7.56 ± 0.07	252	1.92	1.62
5	0	9.59 ± 0.11	401	2.35	0.58
10	0	11.32 ± 0.10	481	2.06	0.52
0	5	7.85 ± 0.09	290	3.02	0.72
0	10	8.42 ± 0.08	303	1.62	0.66
5	5	9.71 ± 0.05	363	2.84	0.47
10	5	8.47 ± 0.08	391	2.26	0.08
5	10	8.93 ± 0.07	407	2.54	0.21

7.2.5 γ- radiation studies

The hybrid filler reinforced EPDM-CIIR blends for product application were irradiated at three different cumulative doses of gamma rays from Co^{60} source.

7.2.5.1 FTIR spectra of γ-Irradiated hybrid filler reinforced EPDM/CIIR blends

FTIR spectra were carried out for all irradiated EPDM/CIIR hybrid samples. For illustration, FTIR spectra of irradiated (1MGy and 2MGy) samples are reported in Table 7.2. No significant changes were noticed in spectra of samples irradiated to low dose of gamma radiation (0.5MGy). The formation of new chemical functional groups corresponding to gamma radiation induced crosslinking and/or degradation for hybrid filler reinforced EPDM/CIIR blends are reported in Table 7.2. The peaks corresponding to wavenumbers 1365 cm^{-1} and 1454 cm^{-1} denotes formation of C-C bonds (stretching), indicating dominance in radiation induced crosslinking for hybrid nanocomposite irradiated at 1MGy. For the hybrid nanocomposites exposed to 2MGy (highest dose) of gamma radiation, intense bands were observed at wavenumbers 3305 cm^{-1}, 3604 cm^{-1}

and 1135 cm^{-1} corresponding to C-O bonds (stretch) and -OH bonds. These oxygenated bonds are formed due to the dominance of chain scission or degradation of the polymer chains. At higher dose of gamma radiation (2 MGy), formation of oxygenated chemical bonds corresponds to dominance of chain scission or oxygenative degradation.

Table 7.2 Chemical groups from FTIR spectra of irradiated EPDM/CIIR hybrid nanocomposites

Sample	Dose of Radiation	Wavenumber (cm^{-1})	Bond	Inference
EP80CI20NS5	Large (2 MGy)	3305, 3604, 1135	-OH, CO	Free radical by chain scission
EP80CI20NC5		3603, 3303 (Reduced)	-OH	Resistant against γ irradiation
EP80CI20NC10		2850, 2915	CH	
EP80CI20NS5	Medium (1 MGy)	1365, 1454	C-C	Occurrence of cross linking
EP80CI20NC5		1030, 1365	Si-O CN	No -OH formation
EP80CI20NC10		2952, 2850 (Intensity increased)	CH	Resistant towards medium dose

7.2.5.2 Mechanical behavior of γ-Irradiated EPDM/CIIR hybrid nanocomposites

The mechanical behavior of irradiated EPDM/CIIR hybrid nanocomposites is estimated upon percentage change in the properties from unirradiated ones as discussed in previous chapters.

The percentage change in mechanical (tensile) strength of irradiated hybrid composites is tabulated in Table 7.3. The least change in properties after irradiation was observed on EP80CI20NC5NS10. The nanoclay acts as anti-rads by scavenging free radicals produced after irradiation. They also induce tortuous path for the free radicals to traverse through the nanocomposite and react with polymer chains as explained in Chapter 4. The modified nanosilica particles also establish covalent linkages with high bond dissociation energy in the blend. These linkages also resist the changes caused by gamma radiation. The combination of NC (5 phr) and NS (10 phr) (hybrid fillers) in CB reinforced EPDM/CIIR blends restricts the changes in structure of the elastomer hybrid composite after gamma irradiation primarily due to synergism between the fillers [277] [175], [295], [296]. The elastomeric components should exhibit minimal change in properties after irradiation for attaining longevity in its utility for nuclear plants.

Table 7.3 Percentage change in tensile strength of irradiated binary and ternary (hybrid) nanocomposites

Nanoclay	Nanosilica	Low Dose (0.5MGy)	Medium Dose (1MGy)	High Dose (2MGy)
0	0	30.1	52.5	-53.9
5	0	14.4	30.8	-35.9
10	0	13.7	43.7	-47.8
0	5	27.5	36.9	-43.1
0	10	9.86	33.7	-52.2
5	5	18.3	26.9	-25.5
10	5	19.5	30.0	-33.7
5	10	10.4	16.2	-27.1

7.2.5.3 Crosslinking density of γ-Irradiated EPDM/CIIR hybrid nanocomposites

The dominance of radiation induced chain scission or cleavage of elastomer chains provides ease of solvent permeation. The prevalence of radiation induced crosslinking imparts barrier to solvent permeation. Crosslinking density increases when more number of crosslinks is present in between elastomer chains. The diffusion experiments were carried out for all irradiated samples. The swelling coeffients and crosslinking density of irradiated hybrid nanocomposites are tabulated in Table 7.4 below.

Table 7.4 Swelling coefficient and crosslink density of hybrid nanocomposites before and after radiation

Nanofiller (phr)		Before γ-Irradiation			0.5 MGy			1MGy			2MGy		
NC	NS	β (cm^3/g)	M_c (g/mol)	$v \times 10^3$ $(gmol/cm^3)$	β (cm^3/g)	M_c (g/mol)	$v \times 10^3$ $(gmol/cm^3)$	β (cm^3/g)	M_c (g/mol)	$v \times 10^3$ $(gmol/cm^3)$	β (cm^3/g)	M_c (g/mol)	$v \times 10^3$ $(gmol/cm^3)$
0	0	0.61	206	4.85	0.59	204	4.89	0.58	195	5.13	0.62	211	4.74
5	0	0.57	196	5.09	0.52	163	6.12	0.41	136	7.33	0.48	164	6.08
10	0	0.22	80.4	12.4	0.18	65	15.3	0.13	55.2	18.1	0.31	107	9.33
0	5	0.57	197	5.06	0.48	186	5.36	0.36	121	8.26	0.52	179	5.59
0	10	0.49	162	6.18	0.42	146	6.85	0.35	123	8.13	0.59	214	4.67
5	5	0.36	121	8.24	0.33	120	8.34	0.34	114	8.75	0.43	46.0	21.7
10	5	0.12	48.9	20.4	0.11	48.3	20.7	0.10	46.0	21.7	0.11	46.6	21.4

As seen in the previous chapters, the decrease in swelling coefficient and increase in crosslink density till 1 MGy when compared to unirradiated samples was due to dominance of crosslinking as observed from mechanical behavior. Dominance of chain scisson or oxidative degradation at 2MGy declined crosslink density of the hybrid samples.

7.2.5.6 ESR analysis of irradiated hybrid nanocomposites

The radiation induced changes like crosslinking and/or chain scission in elastomeric materials generates free radicals as discussed in previous chapters. ESR analysis was carried out for all hybrid samples. But for an illustration ESR spectrum of EP80CI20NC5NS10 is illustrated in Figure 7.7. ESR spectra of un-irradiated and irradiated (0.5, 1 and 2 MGy) hybrid nanocomposite of EPDM/CIIR blends are represented in Figure 7.7. No significant changes were observed in ESR spectra of irradiated samples.

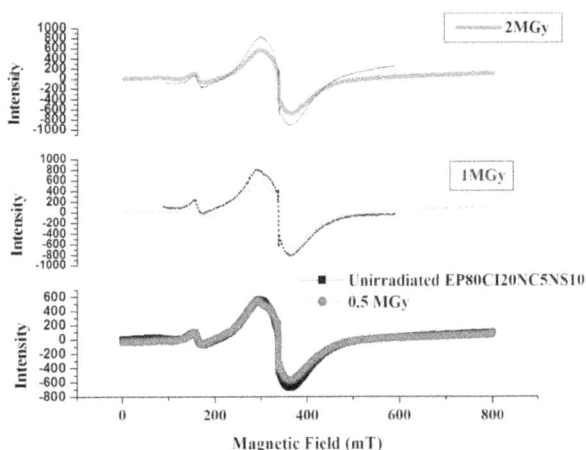

Figure 7.7 ESR spectra of hybrid nanocomposite

7.2.6 Conclusions

EPDM-CIIR hybrid nanocomposites were prepared by two stage process and characterized for mechanical properties, spectroscopic analysis, filler-rubber interactions and gamma radiation ageing behavior. The reinforcement effect was studied from mechanical properties and Payne effect. The uniform and slow decline in storage modulus in composite with well dispersed nanofillers confirmed larger filler-rubber interactions in the matrix. TEM micrographs showed good distribution of nanofillers and few aggregates in the blend with larger filler content. The changes in mechanical properties after gamma irradiation was found to be minimum for 5 phr nanoclay and 10 phr nanosilica because of antirad property of nanoclay and radiation resistance property of nanosilica. ESR spectra revealed absence of free radicals in the structure of ternary nanocomposite. The ternary or hybrid nanocomposites of EPDM-CIIR blends have potential for application in nuclear plants.

Chapter 8

Conclusions and future scope

The studies on nanocomposites of EPDM-CIIR elastomeric blends for gamma radiation environment in nuclear fuel reprocessing facilities led to following major conclusions

- EPDM-CIIR blends were prepared in different compositions and evaluated for compatibility, cure kinetics, mechanical and solvent barrier properties. The synergistic behavior of blends enhanced mechanical behavior and results have been corroborated with morphology and spectroscopic analysis. The blend containing 80% EPDM and 20% CIIR was found to be optimal based on percentage retention in mechanical properties and solvent coefficient after gamma-irradiation.

- Nanoclay based nanocomposites of EPDM/CIIR blends were prepared by binary stage process and evaluated for morphology, mechanical and viscoelastic, rheometric, nano reinforcement mechanism, transport, thermal characteristics and gamma radiation aging behavior. Morphology studies of the nanocomposites showed that the layered silicates were well dispersed in the elastomer matrix with intercalation and exfoliation till 5 phr, though a few aggregates of layered silicates were observed at higher nanoclay level. The TS increased by 59% for blend with 5 phr NC owing to interactions and dispersion of nanoclay.

- Exploration of dynamic mechanical properties showed noteworthy enhancement in storage modulus in the nanocomposites below the glass-transition temperature. The reinforcing parameters were evaluated from DMA and experimentally obtained elastic modulus was compared with those predicted from analytical models. The least change in tensile and solvent properties after irradiation were found for blend with 5 phr nanoclay.

- Influence of bis(3-triethoxysilylpropyl)tetrasulfide (TESPT) modified nanosilica reinforcement on mechanical, viscoelastic, thermal and transport characteristics as well as behavior after exposure to different cumulative doses of gamma rays were studied. The formation of chemically covalent interfaces after silanization was evaluated from FTIR spectra. The changes in surface chemistry of modified NS particles ensured distribution and interactions within the rubber chains. TEM analysis and FTIR spectroscopy corroborated dispersion and chemical interactions of nanosilica in the elastomeric blend. The well dispersed silica particles in blend with 7.5phr enhanced TS and M100 by 43% and 102% respectively.

- The applicability of analytical models for solvent sorption and thermal degradation kinetic behavior were employed in this study. The radiation ageing resistances were improved after incorporating modified nanosilica in the blend. The ability of well distributed nanoparticles in the matrix at 7.5 phr, restricted mobility of free radical transport by providing barrier and chemical interfaces with the rubber restricted the radiation ageing effects. The dominance of crosslinking was observed till 1 MGy, whereas chain scission prevailed at 2MGy. The chemical changes associated with radiation induced changes were observed from FTIR spectroscopy. ESR analysis gave an insight to the presence of free radicals for nanocomposite irradiated at 2 MGy.

- MWCNT reinforced (0.5, 1, 1.5 and 2 phr) blends prepared were evaluated for cure, mechanical, solvent transport and radiation ageing characteristics. The dispersion and chemical interactions were analyzed by FTIR and TEM analysis. The interfacial interactions and effective stress transfer between the nanofiller and matrix resulted in enhancement of mechanical properties with increasing MWCNT content till 1.5 phr MWCNT. Solvent transport coefficients were found to decrease with filler content owing to induction of tortuous path and reduction in free volume by the nano reinforcements. The slight reduction in mechanical properties of blends at higher content of MWCNT may be attributed to agglomeration. MWCNT act as free radical scavenger which when reinforced in blends produced minimized damage caused due to γ-radiation when

165

compared with neat blends. The least change in properties after irradiation were observed in blend with 1.5 phr MWCNT

- The nanofiller-carbon black reinforced hybrid EPDM/CIIR blends were characterized for mechanical properties, filler-rubber interactions and γ-radiation ageing behavior for product application. The hybrid nanocomposite with 5 phr NC and 10 phr NS produced least change in mechanical and swelling coefficient after irradiation, due to hybrid effect of fillers.

- The developed hybrid nanocomposites of EPDM-CIIR blends have potential applications in components like Gaskets, O-rings and Manipulator bootings used in nuclear reprocessing facilities. The nanocomposites have improved gamma radiation and hydrocarbon ageing characteristics along with superior mechanical properties and expected to extend the product life of such components.

Scope for future work

- Finite element modeling and non linear stress analysis of nano filler reinforced EPDM blends.
- Viscoelastic modeling of elastomer nanocomposites based on creep and stress relaxation studies.
- Studies on electron beam and neutron irradiation of EDPM based blends for applications in nuclear, space, etc.
- Studies of nanocomposites of EPDM with other elastomers for various.
- Temperature sweep studies on mechanical and viscoelastic properties of EPDM based nanocomposites.
- Study of filler treatment or modification with different compatibilizers

www.ingramcontent.com/pod-product-compliance
Lightning Source LLC
Chambersburg PA
CBHW071231210326
41597CB00016B/2004